NUTS

BASTOGNE

THE SIEGE OF BASTOGNE, BATTLE OF THE BULGE, DECEMBER 1944

Written by: Wayne Turner

Assistant Writers: Mike Haught, Phil Yates

Editors: Peter Simunovich, John-Paul Brisigotti

Proof Readers: Russell Briant, Joe Gallant, John Goetzinger, Paul Kitchen, Gary Martin, Gregg Siter, Neal Smith

Graphic Design: Victor Pesch

Miniatures Design: Evan Allen, James Brown, Tim Adcock

Cover Art: Vincent Wai

Internal Art: Ben Wootten, Vincent Wai

Miniatures Painting: James Brown, Blake Coster, Mark Hazell Aaron Mathie, Matt Parkes, Shelby Taylor

Terrain Modelling and Photography: Battlefront Studio

Playtest Groups: Arizona Desert Rats (Thomas Weller), Dad's Army (Andrew Duncan), Game Korps Dog Company (Troy Fuller), Oregon Flames Of War (Michael McSwiney), The Regiment (Simon McBeth), Sydney FOW Synicate (David Robinson).

CONTENTS

This is a supplement for *Flames Of War, the World War II miniatures game.*
A copy of the rulebook for *Flames Of War* is necessary to fully use the contents of this book.

BASTOGNE
AND THE BATTLE OF THE BULGE

The Germans launched *Unternehmen Wacht am Rhein* (Operation Watch on the Rhine, also known as the Ardennes offensive) on 16 December 1944. Their aim was to punch through the thinly held lines of the US First Army in the Ardennes region of Belgium and Luxembourg with their infantry, before unleashing the Panzer divisions to drive through to the Meuse River. The overall German goal was to take the Belgian port of Antwerp, so vital for Allied supplies, and cut off the US Ninth Army, British Second Army and Canadian First Army in northern Belgium and the southern Netherlands. The key to the success of the German plan was to quickly cut through the US First Army in the Ardennes.

The main thrust towards Antwerp was given to the powerful *6. Panzerarmee* (6th Tank Army) on the northern shoulder of the offensive. However, determined defence by the US 1st and 99th Infantry Divisions halted the thrust. In the central sector, *5. Panzerarmee* was more successful against the worn and stretched US 28th and inexperienced 106th Infantry Divisions. The vital road junctions at the small town of Bastogne became the focal point of the offensive. All seven of the main roads through the Ardennes converged at Bastogne and the capture of the town would ensure the continued drive by *5. Panzerarmee* towards its objectives.

101st Airborne Division

At the start of the Germans' Ardennes offensive, the 101st Airborne Division was in camp at Mourmelon, France resting and refitting after the hard fought Market Garden campaign and following battles in the Netherlands. On 17 December the division was put on alert to go to the front and on 18 December they were hurriedly loaded into trucks and shipped off to Bastogne.

On arriving in Bastogne on 19 December the first units of the 101st Airborne were thrust into the line to halt the advancing Germans. They were joined by Combat Command B of the 10th Armored Division and the 705th Tank Destroyer Battalion. As they set up a defensive perimeter around Bastogne, they were attacked by *2. Panzerdivision*, *Panzer Lehr Division*, and *26. Volksgrenadierdivision*. Battles were fought at Wardin, Noville, and Foy (see map on page 6) as the paratroopers pushed out from Bastogne on that first day.

By 20 December the 101st Airborne Division had established a defensive perimeter around Bastogne with its four infantry regiments encircling the town.

THE ARDENNES

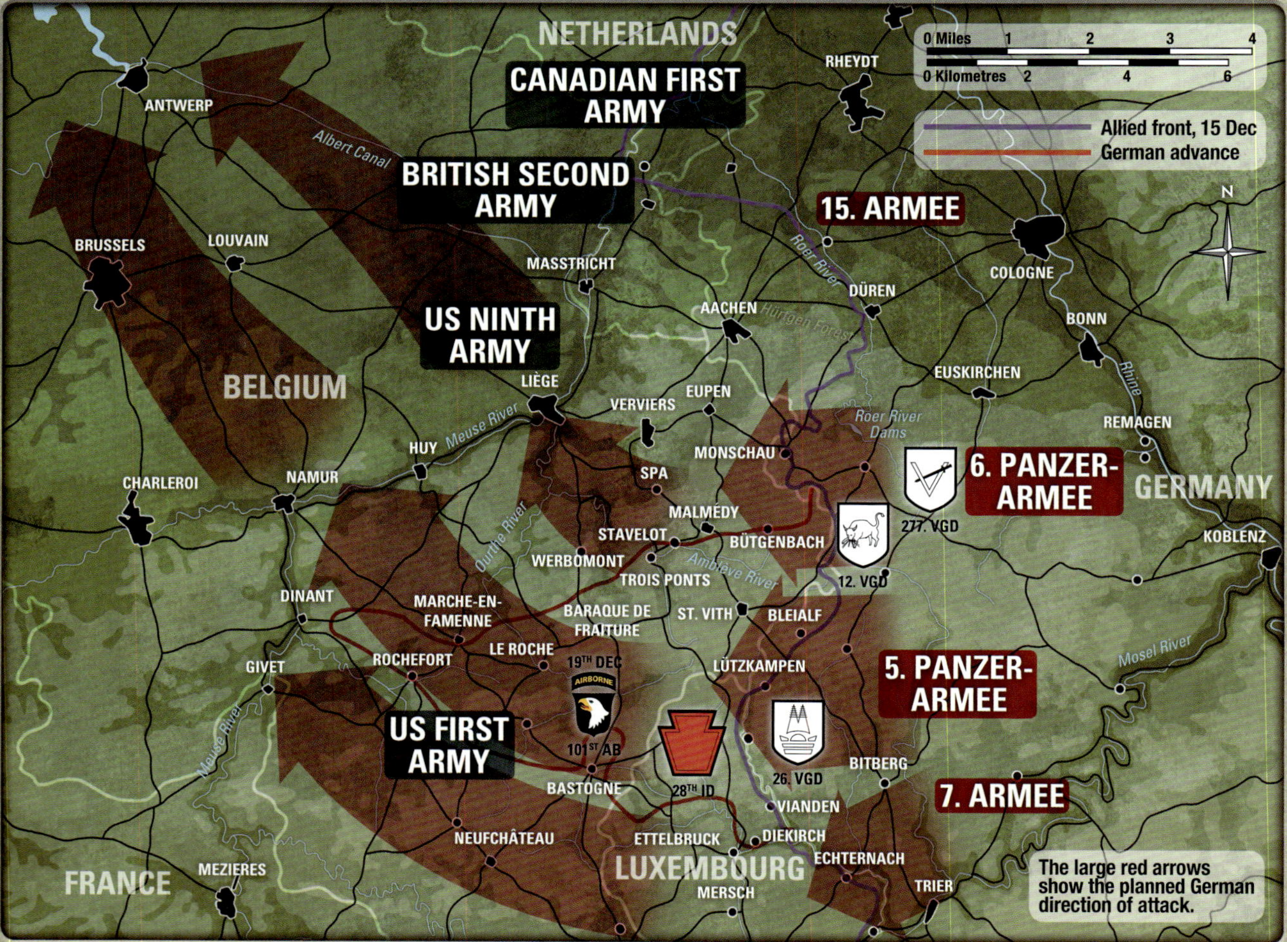

THE SIEGE

With a strong perimeter established, particularly in the north and east, the German *XLVII Panzerkorps* commander, General von Lüttwitz, decided to circle around Bastogne and strike a blow from the south and southwest on the night of 20/21 December with his two panzer divisions. The Germans initially penetrated into the lines in the southwest, but an ad-hoc American counterattack force threw back the attack. However, all the roads in and out of Bastogne had been cut by the Germans and the town was surrounded.

With Bastogne surrounded, the Panzer divisions continued on towards the Meuse River, leaving behind a regiment to support the *26. Volksgrenadierdivision* in capturing Bastogne. *26. Volksgrenadierdivision* and its support probed the southern and western sectors of the American perimeter. Because the Germans didn't launch a concerted attack against several sectors at once, the 101st Airborne Division was able to concentrate its supporting armour and artillery against each attack as it came, moving them about as they were needed.

Continuous attacks had exhausted *26. Volksgrenadierdivision* by 23 December. On 24 December they received a much needed boost in the form of a panzergrenadier regiment from *15. Panzergrenadierdivision.* The Germans then attacked the positions of the 327th Glider Infantry Regiment on the western section of the perimeter on Christmas Day. Initially the attack was successful and the panzergrenadiers and tanks got as far as Hemroulle, where the 327th Regiment had its 3rd Battalion (formally the 1st Battalion, 401st Glider Infantry Regiment) command post. The German panzers leading the assault split into two columns and continued on. One column was immediately ambushed by two M18 Hellcats of the 705th Tank Destroyer Battalion before Champs. The other column headed for Rolle, where it was hit by another two M18s. The German assault had been brought to a halt as the German Panzer IV tanks were knocked out by tank destroyer and bazooka fire and the German infantry assaults were held off by the glidermen.

PATTON BREAKS THROUGH TO BASTOGNE

While the paratroopers and glidermen were holding tenaciously to their positions around Bastogne, General Patton's Third Army, spearheaded by the 4th Armored Division's 37th Tank Battalion (commanded by Lieutenant-Colonel Creighton Abrams), had finally broken through to Bastogne. They reached the lines of the 326th Airborne Engineer Battalion on 26 December. Communications and supplies were quickly restored.

Without the benefit of rest, the 101st Airborne Division, immediately joined the attack. After many more days of hard fighting they finally pushed the Germans back. On 17 January the 101st Airborne had reached the positions they had occupied on 19 December, when the division had first arrived at Bastogne. The following day the 101st Airborne Division was relieved.

THE ALLIES PUSH BACK THE GERMANS

The German offensive had ground to a halt, but had pushed a large salient into the Allied lines. To destroy the salient, Patton's Third Army attacked north, while Montgomery's American and British forces, to the north, struck south in early January. The two forces planned to meet at Houffalize.

Progress was slow in the winter conditions and the majority of the German forces were able to escape the salient. The fuel situation had become so bad for the Germans that most of their panzers had to be abandoned. St Vith wasn't recaptured by the Americans until 23 January and the last German units participating in the offensive hadn't returned to their start lines until 25 January.

Hitler's ambitious plan to take Antwerp and destroy three armies had failed and in doing so gave the Americans one of their most triumphant victories of the war.

BASTOGNE

Volksgrenadiers and their supporting Hetzer tank-hunters assault a US position in an attempt to break into Bastogne.

The glidermen and paratroopers made good use of Bazookas against German panzers such as this tough Panther.

Defensive positions were set up all around Bastogne by the 101st Airborne as they fought to hold off German attacks.

The elite infantry of 101st Airborne Division were supported by Sherman tanks and M18 Hellcat tank destroyers.

The German forces attacking Bastogne were well supported with armour, keeping the tank destroyers busy.

SCREAMING EAGLES

101ST AIRBORNE DIVISION IN BASTOGNE

The 101st Airborne Division was formed at Camp Claiborne, Louisiana on 15 August 1942. By the time the division saw its first combat on 6 June 1944 the men of the 'Screaming Eagles', named for the emblem on their shoulder patch, had undergone extensive and intense training in the United States and United Kingdom. Having acquitted themselves well and played an important role in securing the Normandy beachheads, the division was next dropped on the Netherlands during Operation Market Garden. Once again they fought with distinction, securing 17 Miles (28km) of road stretching from Eindhoven to Grave Bridge, keeping 'Hell's Highway', as it became known, open for the British troops of 30 Corps.

101ST AIRBORNE HEAD FOR BASTOGNE

As the size and success of the initial attacks by the German divisions in the Ardennes became clear, the 101st Airborne Division was put on alert at their camp in Mourmelon, France on 17 December 1944. The division quickly set about calling men back from leave, including a number of commanders, as well as organising the re-issuing of arms and ammunition. Transport was hastily arranged and the paratroopers and glidermen were loaded onto trucks on 18 December and moved off to the front.

How prepared each unit of the 101st Airborne Division was depended greatly on the initiative of the commander and the men. Not every man was able to grab their weapons or enough ammunition.

Lieutenant-Colonel John T. Cooper of the 463rd Parachute Field Artillery Battalion had the forethought to redirect his part of the column through the ammunition dump at Mourmelon, where he got his men to grab as much ammunition for their 75mm pack howitzers as they could carry. The 63rd Parachute Field Artillery Battalion Italy Veterans temporarily attached to the division while waiting for the 17th Airborne to arrive in France. Other units were less well prepared and took weapons and ammunition off US riflemen as they encountered them retreating from Bastogne. Ammunition was also supplied by the 10th Armored Division's Combat Command B (CCB), which had arrived at the front the day before the airborne division.

As well as an initial shortage of weapons and ammunition, the division was also without their commander, Major-General Maxwell Taylor, who was away in Paris. Instead, the division was commanded by General Anthony McAuliffe, the division's second in command.

BASTOGNE 19 TO 25 DECEMBER

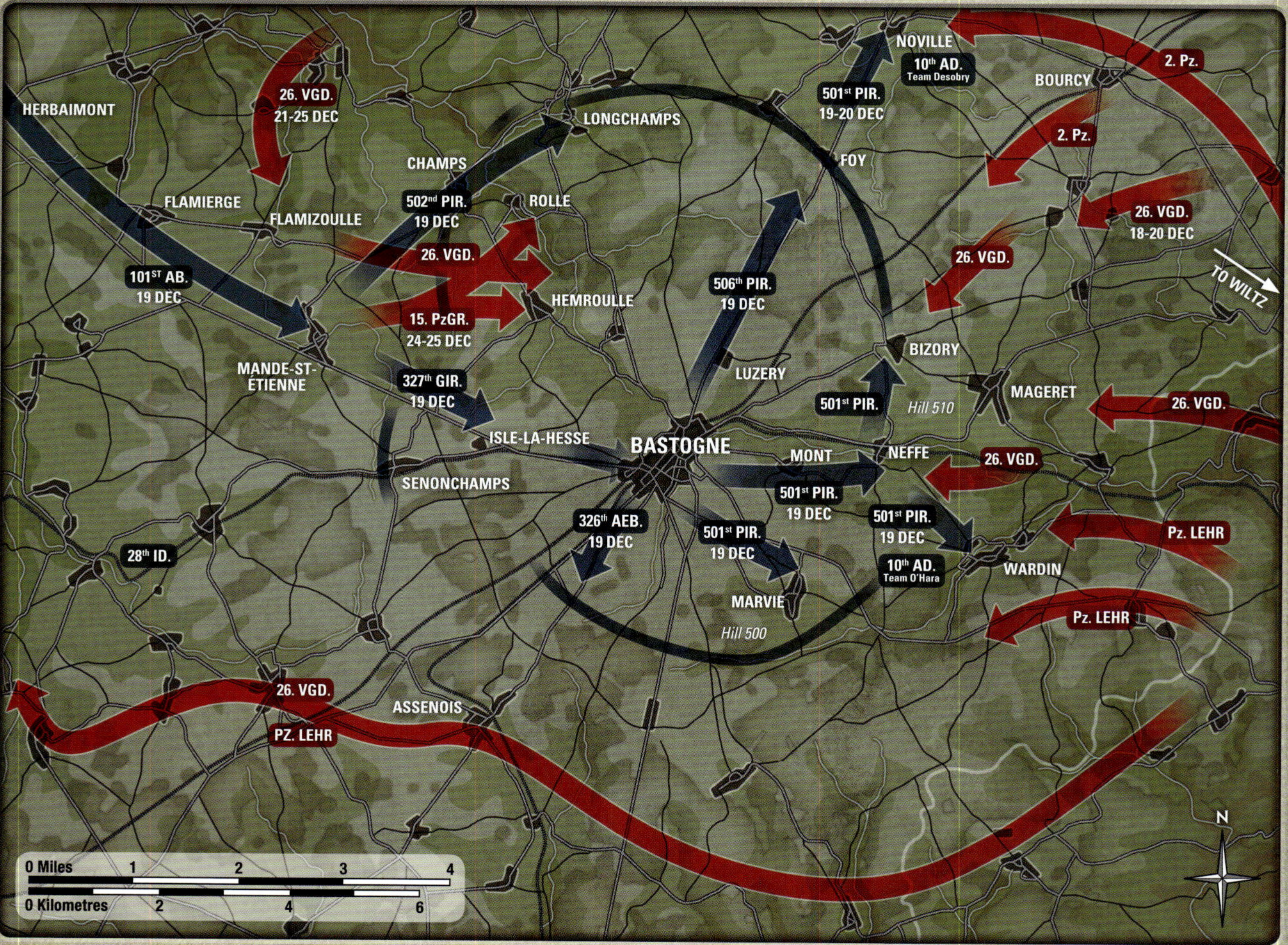

The Division arrives in Bastogne

After driving through the night, the division arrived at Bastogne on 19 December 1944. As the 101st Airborne Division arrived, the units were pushed forward to establish the extent of the German advance and to set up a perimeter around the town. The first to arrive at the front were the men of the 501st Parachute Infantry Regiment (501st PIR) and they were soon sent forward through the fog to 'go out and find the enemy'. The 1st Battalion moved out first and expected to find the enemy near the village of Margeret. As they advanced, they met scattered groups of men from the 28th Infantry Division retreating back towards Bastogne.

Leading them forward were the jeep-mounted men of the divisional recon. As they rounded a corner near the village of Neffe, all hell broke loose. The recon jeeps had been ambushed by a German tank and some infantry. Following behind the recon was B Company, 1st Battalion. They quickly sent forward some bazooka men and let off a volley towards the tank, which halted and was then eventually knocked out. B Company and the battalion headquarters behind them quickly spread out and deployed in defensive positions.

At 1000 hours, the rest of the regiment left Bastogne to join the 1st Battalion near Neffe. 2nd Battalion moved just north of Neffe to Bizory with the aim of taking Mageret from that direction. They were halted quickly by the German *26. Aufklärungs Abteilung, 26. Volksgrenadierdivision* who had already entrenched on top of Hill 510 overlooking the road from Bizory. However, when the Volksgrenadiers continued their advance, they were caught in an artillery barrage from the 907th Glider Field Artillery Battalion, forcing the Germans back to their positions on Hill 510.

Wardin

The 3rd Battalion, 501st PIR advanced towards Mont, southwest of Neffe, but didn't arrive until after mid-day due to traffic in Bastogne. The 101st Airborne Division became tangled with other US units retreating from the German onslaught. Once the battalion had arrived in Mont, I Company, the battalion's southern flank guard, was sent forward to check the three wooded areas before Wardin. Once reported clear, the company was then ordered to move into Wardin and make contact with a supposed US roadblock established there. However, as I Company entered Wardin, troops of Team O'Hara, CCB, 10th Armored Division could be seen moving towards Bastogne along a ridgeline to the southwest. The roadblock troops had left.

The reason soon became clear as a force of tanks and infantry from the German *Panzer Lehr* Division's *901. Panzergrenadier-regiment* began to attack Wardin. Two German tanks knocked out a Team O'Hara tank as they moved into the village. I Company managed to knock out two tanks with their bazookas before the Germans overran the village and forced the company to retreat back towards Mont.

As the rest of 3rd Battalion moved into Mont, they directed patrols toward Neffe, which they soon discovered was well and truly infested with Germans. As G Company probed towards Neffe, they joined with Team 'Cherry' of the 10th Armored Division who had been pushed out of Neffe. The battalion decided to pull back and set up their defence at Mont.

To anchor the 501st PIR's right flank, A and C Companies, 1st Battalion, 327th Glider Infantry Regiment (327th GIR) were sent forward to join the paratroopers.

Noville

Team Desobry, CCB, 10th Armored Division had been holding the village of Noville, northeast of Bastogne, since the evening of 19 December collecting stragglers from the 28th Infantry Division and 9th Armored Division to help hold the positions.

Shortly after daybreak on 19 December Major William Desobry was told his team would soon be joined by the 1st Battalion, 506th Parachute Infantry Regiment (506th PIR) at Noville. Desobry had been concerned that his positions at Noville were overlooked by high ground on three sides and had wanted to withdraw. With this news, he sent a jeep to pick up the parachute battalion's commander, Lieutenant-Colonel James LaPrade, so they could work out where to position the paratroopers when they arrived. LaPrade informed Desobry that many of his men were short of weapons and ammunition. Desobry quickly organised equipment to be placed along the road to Noville so the paratroopers could arm themselves as they arrived.

On arriving at about 1400 hours, the companies of the battalion set about securing the high ground around Noville. B Company was to take the high ground on the northeastern flank, C Company the wooded ridgeline on the eastern flank, and A Company was positioned between them in reserve. Preceded by artillery fire and supported by the tanks of Team Desobry, the paratroopers attacked the high-ground. The fog periodically lifted, allowing LaPrada and Desobry, observing from Noville, to see the progress of the attack. Desobry was particularly impressed by the aggressive and speedy advance of the paratroopers.

As the paratroopers of B Company reached the top of the heights, they ran into a German attack led by the tanks of *3. Panzerregiment, 2. Panzerdivision.* The Germans had chosen that very moment to launch their attack on Noville. After two hours of fighting, the two commanders decided it would be better to defend from the buildings of Noville and ordered the paratroopers to withdraw. The fog continued to play an important role, allowing the paratroopers to conduct their fighting withdrawal with some concealment, as well as letting the M18 Hellcat tank destroyers of Team Desobry pick off the panzers as the fog regularly rose and fell.

As the troops withdrew into Noville, LaPrade called a meeting of his company commanders. Just after the meeting had broken up, an enemy artillery shell hit the building housing the battalion command post, killing Lieutenant-Colonel LaPrade and seriously wounding Major Desobry. However, the battalion's executive officer, Major Robert F. Harwick, had just arrived in Noville, after making his way on foot from Bastogne, and was able to take command.

The battle raged around Noville as the Germans continued to pound the village with artillery. Panzers continued to probe forward, but the tank destroyers kept them at bay. The paratroopers and armoured troops still held the village as night fell.

The 506th PIR deployed around Foy, southwest of Noville, on its arrival in Bastogne and dug-in on the high ground. The 502nd Parachute Infantry Regiment (502nd PIR) deployed behind the 506th PIR, north of Bastogne around Champs. The 327th GIR covered the northwestern approaches to Bastogne around Mande St. Etienne, while the 326th Airborne Engineer Battalion covered the south. As the men set up their defensive perimeters and settled in for the night, they could hear the sound of fighting off to the east.

During the evening, the 101st Airborne Division suffered its first major blow when their Field Hospital at Herbaimont was captured by German troops circling around Bastogne. Among those captured was Major Desobry of CCB, 10th Armored Division.

Second Day of Fighting in Noville

As 20 December dawned, with the ever-present fog, the Germans continued the assault on Noville. Initially, they sent their tanks against the paratroopers of the 1st Battalion, 506th PIR in Noville. The fire of the paratroopers along with the men of Team Desobry, and supporting fire of the 321st Glider Field Artillery Battalion, kept the Germans at bay.

The Germans then split into two forces, one headed north around Noville, while the second headed south. The southern probe was halted by the 2nd Battalion, 506th PIR at Foy and the 2nd Battalion, 501st PIR south of them. The German thrust to the north was more successful. By noon, German patrols had started to cut the road between Noville and Foy. The 1st Battalion, 506th PIR had also lost direct contact with the regimental HQ and was relaying messages through the artillery observer's radio link with the 321st Glider Field Artillery Battalion. They were in danger of being cut off. The situation had also become critical in the front line. With casualties mounting, radio operators and clerks from the HQ were sent to the companies to plug the gaps.

The order finally arrived from the regimental HQ to withdraw at 1315 hours. The German attack to the south had begun to gain ground, further threatening to isolate the battalion at Noville. Major Harwick ordered the withdrawal at 1330 hours. C Company and four Sherman tanks from Team Desobry, some crewed by paratroopers, were to move back first to the high ground to the east of Foy to cover the withdrawal of the rest of the force. The rear guard was formed by the tank destroyers and A Company. 3rd Battalion, 506th PIR would counterattack Foy, which was still largely in German hands, to cover the withdrawal.

The US column left Noville behind, with the wounded carried on half-tracks, jeeps and trucks, and the paratroopers and armored infantry on foot beside them. It wasn't long before C Company ran into German roadblocks. The Germans had taken up positions along the road in houses, barns and woods. Fighting flared up all along the road as the column encountered German positions. As the fire from the Germans intensified, the paratroopers moved off to the west side of the road to utilise a natural depression where the Germans couldn't see them. Eventually, as the 3rd Battalion attacked in the opposite direction, the 1st Battalion finally made it back to their lines. Of the 600 men who had entered Noville, less than 400 made it back to Foy.

Establishing the Front Line

The US positions near Foy had also been under attack from *2. Panzerdivision* during the day, which the 3rd Battalion,

506th PIR held until ordered to the counterattack in the afternoon. The 2nd Battalion, 506th PIR had watched from the south of Foy, too far away to help, as the other two battalions struggled to take Foy. However, they had concerns of their own, trying to maintain contact with the 2nd Battalion, 501st PIR to their south. Contact between the two battalions had not been consistent since both had arrived in their positions. Captain Richard Winters, in acting command of the 2nd Battalion, 506th PIR was determined to make sure the Germans couldn't penetrate along the railway line that was the boundary line between the two regiments. He sent F Company to clear the Germans from the gap and make contact, but the German machine-gun fire halted the paratroopers' advance.

The command of the 501st PIR was also concerned about the gap and sent A Company, 1st Battalion along the railway line at about 1500 hours to clear the gap of any enemy and take up positions linking the two regiments. At dusk they made contact with German Volksgrenadiers attempting to filter through the gap and a firefight broke out. The company became pinned down along the railway line and eventually pulled back to higher ground with the fall of darkness.

The South and West

In the small hours of 20 December the 2nd Battalion, 327th GIR was sent to Marvie to take up defence there. A roadblock had been set up by Team O'Hara, CCB, 10th Armored Division and had held off a push down the Wiltz-Bastogne Road by German troops. As the glidermen began setting up in Marvie, they were attacked by a force of German tanks and infantry who had bypassed the armoured troops. The Germans were eventually halted, but maintained a foothold in the village, occupying some buildings at its edge. The 1st Battalion, now reinforced by the arrival of a third company, continued to defend positions around Neffe.

At Neffe and Mont, the paratroopers of the 501st PIR had to withstand a German night attack at about 1900 hours on 20 December. The paratroopers raked the advancing panzergrenadiers of *Panzer Lehr Division* as they became tangled in the barbed wire cattle fences crossing the fields before Mont. M18 Hellcat tank destroyers took care of the Germans' supporting assault guns.

21 December

On 21 December the glidermen of 3rd Battalion, 327th GIR continued to patrol aggressively in the west, as the Germans began to fully encircle Bastogne. The weather had also made a turn for the worse. After several days of moderate winter conditions, temperatures had dropped sharply and it began to snow. The haste of the airborne troops' departure from their camp in France began to take its toll. Many men didn't have greatcoats or overshoes and their only protection from the cold and snow were shelter quarters, sleeping bags, and blankets.

The glidermen and engineers, with support from the CCR, 9th Armored Division continued to repel attacks from the west and south throughout the day.

Railway Line Gap

There was to be no rest for the 1st Battalion, 506th PIR after their two-day fight at Noville. During the night, the Germans had moved more troops into Foy, as well as a group of about 200 troops that had penetrated the perimeter along the railway tracks and pushed into the woods not far from the regiment's command post. The battalion was called on to push them out. At 0900 hours, A and C Companies set out along the road towards Foy from Luzery. They then turned south into the woods just south of the railway station and began probing for the enemy positions. They soon found the well-concealed German position and a firefight broke out. Both companies attacked and soon had the Germans on the run, many of whom ran right into the 501st PIR companies across the railway line, providing the anvil for 506th PIR's hammer. By 1600 hours that afternoon, all the infiltrating Germans had been accounted for and the gap between the 506th and 501st Parachute Infantry Regiments had finally been cleared. The paratroopers had killed 65 and captured 165 men of the 1st Battalion, *77. Volksgrenadierregiment, 26. Volksgrenadierdivision.*

22 December

While the fighting had settled down in most sectors, the glidermen of A and C Companies, 3rd Battalion, 327th GIR would spend most of the day in combat. Their fighting centred around Mande St. Etienne where the Germans had

cut the road to Bastogne. After a hard-fought assault, they were able to clear the roadblock. The unit was personally congratulated by General McAuliffe as he toured the perimeter shortly after the battle.

Shortages of food and ammunition became evident throughout the day, as did a lack of white camouflage clothing. The olive green of the US uniforms made a stark contrast against the white snow covered terrain. All around the 101st Airborne Division's front, patrols were out watching the Germans and waiting for them to make their next move.

Nuts!

At 1100 hours a delegation of Germans with a white flag approached the lines of the 327th GIR. They were met by Staff Sergeant Carl Dickinson of F Company. Once the Germans explained they wanted to deliver an ultimatum, they were blindfolded and taken to the company command post where their message was sent on to General McAuliffe in Bastogne.

General McAuliffe was in a deep sleep when Lieutenant-Colonel Ned D. Moore shook him awake to tell him a German delegation had come to take the division's surrender. Groggy from being awakened suddenly, the General muttered 'Nuts' and crawled out of his sleeping bag. After talking over their response it was decided 'Nuts' made an excellent official answer, was typed up and passed on to F Company. After it was explained to the Germans that the answer meant 'go to hell'; they were escorted back to their lines.

The ultimatum called for the division to surrender by 1600 hours or the Germans would level Bastogne with heavy artillery. It turned out to be a somewhat hollow threat, as after the delivery of General McAuliffe's response the shelling did not increase.

23 December

The following day, the Germans did increase their pressure on the positions of the 3rd Battalion, 327th GIR (401st). German infantry and tanks attacked their positions west of Mande St. Etienne in the early hours. Determined defence by the glidermen, timely mortar fire and the support of tanks and tank destroyers kept the Germans at bay. However, by nightfall, some positions had been overrun and the battalion was forced to withdraw their companies. Paratroopers and airborne engineers were also heavily engaged around Marvie in the south, forcing the men on Hill 500 to withdraw.

Requests for resupply resulted in the landing of two sticks of pathfinders on 23 December to set up the landing zone for aerial supply drops. Around 230 C-47 aircraft loads of supplies were successfully dropped on or around Bastogne, a welcome Christmas present for the besieged airborne troops.

Christmas Eve

The Germans gave no respite for festivities and continued to attack in the Marvie sector into 24 December. None of the German attacks were successful in dislodging the airborne troops. Around the rest of the Bastogne perimeter things were quiet, especially compared to the previous few days, broken only by the odd burst of artillery or machine-gun fire. However, during the night, Bastogne was bombed twice and the 10th Armored Division aid station was hit.

Christmas Day

In the west, 3rd Battalion, 327th GIR (401st) was deployed around Hemroulle, 1.5 miles (2.4km) northwest of Bastogne. Their reconnaissance of the German positions indicated the Germans were building up for an attack. The glidermen's reconnaissance was good. *15. Panzergrenadierdivision* had just been committed to the battle and were about to attack through their positions towards Bastogne.

A column of German tanks and assault guns carrying panzergrenadiers advanced into the lines of the battalion. The M18 Hellcat tank destroyer crews of the 705th Tank Destroyer Battalion let the German tanks pass. The men in the defensive line waited patiently while the panzers and riders advanced ahead of more panzergrenadiers on foot following behind. The German panzer column was being followed by the tank destroyers as they approached the glider battalion's command post. It was just before dawn when the glider infantry opened up on the German panzergrenadiers in an open field before their positions. This signalled the tank destroyers to fan out behind the panzers and let loose at their vulnerable rear armour. Five German tanks were quickly destroyed, three by tank destroyers and two by bazookas.

The German tanks then split into two groups. One group of six Panzer IV tanks headed towards Hemroulle and the command posts of the 3rd Battalion, 327th GIR (401st), 1st Battalion, 502nd PIR and 463rd Parachute Field Artillery Battalion, while 12 Panzer IV tanks and StuG assault guns went towards Champs.

Hemroulle

The Hemroulle column ran into the lines of C Company, 502nd PIR. The surprise of the attack had C Company on the run initially, but they rallied, were joined by two tank destroyers, and headed back towards the Germans. They reached Hemroulle about 0800 hours and immediately attacked. Four German tanks headed north, with German infantry scrambling to climb aboard, but the German Panzer IV tanks were quickly knocked out by the tank destroyers.

Another group of German tanks ran afoul of the 463rd Parachute Field Artillery Battalion, who used flanking shots from their little 75mm M1A1 pack howitzers to knock out eight Panzer IV tanks when they came under attack.

Champs

At Champs, A Company, 502nd PIR bore the brunt of stopping the main German thrust towards Bastogne. At about 0300 hours, the Germans had begun to shell Champs with artillery before the infantry of *77. Volksgrenadierregiment, 26. Volksgrenadierdivision* began their attack. Initially, the attack came from the north and the fighting continued until dawn, but the Germans were unable to penetrate the American line. Continued fighting along the front was chaotic with the Germans filtering through the various gaps between A Company's platoons. German tanks also joined the attack, but as their supporting infantry dwindled away, the tanks withdrew.

The enemy then redirected their attack towards Rolle and Longchamps. B Company, 502nd PIR set up a roadblock with one platoon just south of Champs facing the south and the attack coming from the direction of Hemroulle. Supported by two tank destroyers, they were able to knock out five German tanks in the encounter and halt the German advance north, but at the cost both tank destroyers. The rest of the company pushed on to Longchamps.

At dawn, the Germans launched another attack on Champs. The attack, supported by four tanks, was halted by artillery fire, tank destroyers, and the appearance of P-47 fighter-bombers. German infantry soon began to surrender. Though the last of the battles with the German tanks had ended by about 0930 hours, one last tank did make a mid-afternoon appearance from the direction of Bastogne and was knocked out by a bazooka as it headed north through Champs.

A little further south, around Senonchamps, the line was being held by 1st Battalion, 327th GIR, 'Snafu' force, made up of surviving members of the 106th and 28th Infantry Divisions, and some tanks of the 9th Armored Division. They also withstood attacks during the early hours of Christmas Day.

The Siege is Broken

While the 101st Airborne was fighting off the German assaults from the west, the 4th Armored Division was doing its best to break through the German encirclement and relieve the beleaguered airborne troops in Bastogne. On 26 December the Germans continued their pressure on the Bastogne perimeter, attacking in the sectors of the 501st PIR, 327th GIR, 502nd PIR and the 326th Airborne Engineer Battalion. It was a clear day and American P-47s roamed the skies looking for targets, making German attacks that much more difficult.

By the afternoon, the 4th Armored Division was southwest of Bastogne and less than 2 miles (3km) away. Lieutenant-Colonel Creighton Abrams decided to send C Company of his 37th Tank Battalion up the secondary road through Assenois to Bastogne. They punched through Assenois under an artillery barrage, leaving the village to cleared by the following 53rd Armored Infantry Battalion. The direction of their attack took the Germans by surprise and the tanks smashed through the German lines. They soon reached the American line, and after some hesitation, Lieutenant Duane Webster of the 326th Airborne Engineer Battalion came out to meet 2nd Lieutenant Charles Boggess of C Company. After introductions Webster exclaimed, 'Glad to see you!' It wasn't long before Colonel Abrams was in Bastogne meeting General McAuliffe. Patton's Third Army had made it to Bastogne at 1640 hours on 26 December. The siege was over.

4th Armored Division Arrives 26 December

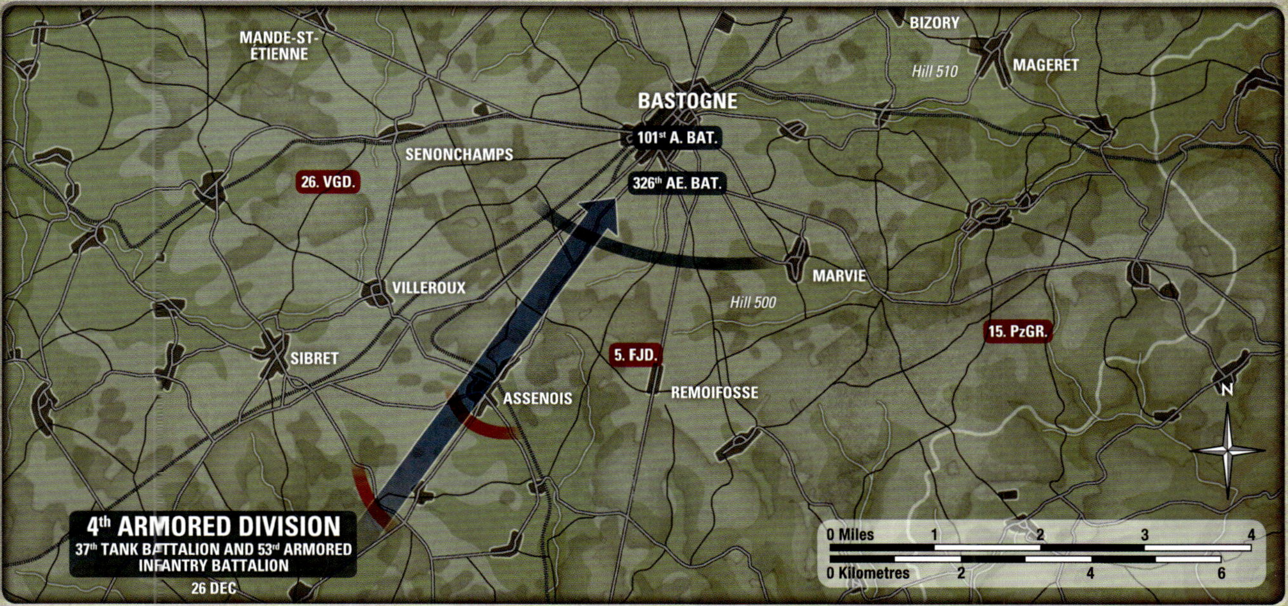

EASY COMPANY

A COMPANY OF HEROES

"'Grandpa, were you a hero in the war?' Grandpa said 'No... but I served in a company of heroes.'"

Dog, Easy, and Fox Companies formed the second battalion of the 506th Parachute Infantry Regiment (PIR) in 1942. The paratroopers were a new experimental arm in the US Army. Under the instruction of Lieutenant Herbert Sobel, Easy Company's original class of 140 recruits were pushed to their limit until only the most tough and focused troops remained. In September 1943, Easy Company arrived in the United Kingdom with the rest of the 101st Airborne Division to prepare for the invasion of France.

NORMANDY

Easy Company landed in Normandy on the night of 5 June 1944 as a part of Operation Overlord. Like the other parachute rifle companies, Easy was scattered across the Normandy countryside during the jump, but quickly regrouped and distinguished itself knocking by out a battery of 105mm howitzers at Brécourt Manor and again during the capture of Carentan.

HOLLAND

Easy Company returned to England and prepared for their next battle. In September 1944, Easy Company jumped into Holland during Operation Market Garden and, along with the rest of the 101st, worked to secure what would be called Hell's Highway. Easy's main battle took place at Nuenen. Working closely with the British 11th Armoured Division, the paratroopers fought the 107th Panzer Brigade, finally ejecting the Germans from the village.

THE ARDENNES

After continued fighting in Holland into November, the 101st was finally withdrawn to Mourmelon-le-Grand to rest and refit. They had barely settled into garrison duty when the call came to mount up on trucks and move out. Leaving with only the ammo and supplies they had brought with them out of Holland, the paratroopers were rushed to Bastogne, Belgium.

Easy took up its position on 2nd Battalion's left flank, just inside the tree line overlooking the village of Foy. The perimeter was far from secure, as Germans and Americans stumbled into each other's foxholes in the snow and fog. At Bastogne, Easy endured relentless artillery bombardments, fought off several German attempts to attack Bastogne, and battled the elements.

Finally, on 9 January, Easy Company launched an attack on Foy itself. Easy launched the attack at 0900 hours, crossing an open field between the woods and the village. The assault stalled in front of the Germans due to the poor leadership of Lieutenant Norman Dike who was paralysed with fear. Dike was quickly replaced by Lieutenant Ronald Speirs who led the attack forward. Easy overcame the Germans in Foy, finally securing the town by the afternoon.

After Foy, Easy joined the rest of the 101st in counterattacking the Germans and pushing them back into Germany and on into Austria, but the fight in the Bastogne woods would be remembered as the hardest and coldest battle fought by Easy Company during the war.

CAPTAIN RICHARD WINTERS

Captain Richard Winters is a Higher Command Rifle team with Gammon Bombs rated **Fearless Veteran**. Winters can be added to a Parachute Rifle Company (page 16) for +70 points.

Richard Winters joined the paratroopers in August 1942 and reported to Easy Company while it was being formed at the Toccoa training facility. Winters quickly earned the trust of the men of Easy Company.

Winters took command of Easy for the first time following the jump into Normandy, after the company commander was killed. He led the company in a bold assault at Brécourt Manor, for which he was awarded the Distinguished Service Cross. In Holland, Winters once again demonstrated his leadership and he was promoted to Captain and, by October, Winters was made the 2nd Battalion's executive officer.

During the Battle of the Bulge Winters kept a close connection with Easy Company. When the hard missions came up he relied on Easy Company to get the job done. At Foy, for example, Winters carefully monitored Easy's assault and quickly replaced Dike with Speirs to press home the attack.

Winters was highly regarded by the men of Easy and, under his leadership, the company remained steady under some of the war's toughest trials.

Captain Nixon

Captain Lewis Nixon was a good friend and confidant of Winters. Nixon worked closely with Winters to provide Easy Company with the best intelligence possible for the mission at hand. Captain Nixon is modelled on Captain Richard Winters's team base.

Platoon Command teams from Combat and Weapons platoons may use the Eyes and Ears rule (found on page 195 of the rulebook) as though they were Recce teams.

A Band of Brothers

Under Winters' leadership, Easy Company pulled together and fought as a cohesive and professional unit throughout the war from France to Germany.

A Parachute Rifle Company led by Winters may re-roll the first Company Morale Check that it is required to take if it fails its first attempt in a game.

Captain Richard Winters and Captain Lewis Nixon.

Lieutenant Ronald Speirs' defiant dash.

LIEUTENANT RONALD SPEIRS

Lieutenant Ronald Speirs replaces the Company Command Carbine team of a Parachute Rifle Company (page 16) for +45 points. He is a Company Command SMG team with Gammon Bombs rated **Fearless Veteran.**

As a Lieutenant, Ronald Speirs joined D (Dog) Company at Toccoa in 1942. In Normandy, he earned a reputation as being a cold-blooded killer. Rumours abounded, where in one case Speirs is said to have gunned down several German prisoners of war. In another instance, Speirs shot one of his own squad leaders for disobeying an order twice which, if ignored, could have resulted in heavy casualties. The speculation only served to bolster his reputation as a fearless and tough fighter.

When Captain Winters saw Easy Company's assault on Foy come to a halt due to Lieutenant Norman Dike's poor leadership, he spun around and ordered the first officer to come into view to go relieve Dike and finish the assault. That man happened to be Speirs, who instantly sprang into action and carried the men to victory through extreme acts of bravery.

After Foy, Speirs officially became the company's longest serving commanding officer during the war and the men took to his leadership well. Speirs remained in command until the end of the war. Winters often praised Speirs as one of the best officers in his battalion.

I'm Here to Take Over

When Lieutenant Norman Dike froze during Easy Company's assault on Foy, Speirs was there to take command and press home the attack.

Do not deploy Speirs when you would normally deploy your Company Command team. Instead, at the beginning of any of your Starting Steps, when Ambushes are normally placed, you may place Speirs within 6"/15cm of any teams of a Parachute Rifle Platoon, and not within 4"/10cm of enemy teams.

Defiant Dash

During the attack on Foy, not only did Speirs dash through the enemy's position to meet up with I Company on the other side, but after linking up with Item Company, he then ran back through the enemy to lead Easy's final assault.

While Speirs is not in Bulletproof Cover, he may re-roll failed Infantry Saves for each Hit he takes during the opponent's Shooting Step.

FIRST SERGEANT CARWOOD LIPTON

Lipton was one of the four original privates assigned to Easy Company when it was first formed. He quickly rose through the ranks to Sergeant and earned the respect of the men of Easy Company.

In Normandy, Lipton linked up with Winters and together they prepared for the assault on Brécourt Manor, during which he earned a Bronze Star. Following the fighting in Normandy, Lipton was promoted to First Sergeant of Easy Company.

During the siege of Bastogne, Lipton battled bad morale as well as the Germans. The company was led by Lieutenant Dike, an indecisive and disliked commander. Lipton concerned himself with the spirit of the men. He listened to and gathered all the complaints and approached Winters with their concerns. Although there was nothing Winters could immediately do about Dike, when the assault on Foy faulted, Winters wasted no time in removing him from command.

Grateful and mindful of Lipton's command during the cauldron of the Bulge, Winters secured a battlefield commission to 2nd Lieutenant for Lipton and he joined the battalion staff.

First Sergeant Carwood Lipton.

First Sergeant Carwood Lipton replaces the 2iC Command Carbine team of a Parachute Rifle Company (page 16) for +25 points. He is a 2iC Command SMG team with Gammon Bombs rated **Fearless Veteran**.

'Talk to your Sergeants'

As First Sergeant, Lipton had his ear to the ground to help address the men's worries and fears.

Combat and Weapons Platoons of a Parachute Rifle Company joined by Lipton may re-roll failed Platoon Morale Checks.

PRIVATE EUGENE ROE

Private Eugene Roe.

Eugene Roe, of Bayou Chene, Louisiana enlisted in 1942 and became Easy Company's medic for the duration of the war. He joined the company in Normandy and Holland, where he was wounded on 17 September.

During the Ardennes, Roe and the other medics were instrumental in keeping Easy Company's strength up for fighting, scrounging medical supplies where they could and seeing to the men's wounds and ailments.

Roe had a particular ability to be wherever he was needed the most. Easy Company's Lieutenant Foley recalled:

"He was there when he was needed and how he got 'there' you often wondered. He never received recognition for his bravery, his heroic servicing of the wounded. I recommended him for a Silver Star after a devastating firefight when his exploits were typically outstanding. Maybe I didn't use the proper words and phrases, perhaps Lieutenant Dike didn't approve, or somewhere along the line it was cast aside. I don't know. I never knew except that if any man who struggled in the snow and the cold, in the many attacks through the open and through the woods, ever deserved such a medal, it was our medic, Gene Roe."

Private Eugene Roe may join a Parachute Rifle Company (page 16) for +50 points. He is not a team, but rather a marker.

Always There When Needed

It seemed amazing to the men of Easy that Roe seemed to be everywhere patching up and evacuating the wounded.

Once per turn, if an Infantry or Man-packed Gun team from one of your Combat or Weapons platoons fails its Save, before removing it from play, you may roll a die:

- *On a result of 5+ Roe patches up the men and the Infantry or Man-packed Gun team ignores the failed Save and fights on as normal.*
- *Otherwise, Roe determines that the men need to be evacuated and the team is Destroyed as normal.*

In either case, place Roe's marker next to that team to remind you that you have used his special rule this turn.

TOCCOA SERGEANTS

Toccoa Sergeants.

Like most units in many armies, the backbone of Easy Company was its platoon commanders. These men were brave and cunning, giving the men encouraging words in their own special ways.

By December 1944, many of the platoon leaders of Easy Company were Toccoa men, that is to say men that had been with the company since the beginning. These Non-commissioned Officers were essential to the survival of Easy Company. Names such as 'Wild Bill' Guarnere, Don Malarkey, Joe Toye, John Martin, Denver 'Bull' Randelman, George Luz, Joseph Liebgott, Warren 'Skip' Muck, Frank Perconte, among others, would bolster Easy Company and see it through the cauldron of Bastogne. Easy Company could also count on experienced officers such as Lynn 'Buck' Compton and Harry Welsh to lead platoons.

Bastogne challenged Easy Company like no other battle, but it was tight leadership and unfailing comradeship that saw it through its darkest hour.

Toccoa Sergeants replace any or all Platoon Command teams from Combat and Weapons Platoons of a Parachute Rifle Company (page 16) for +10 points per Toccoa Sergeant. They are Warrior Command SMG teams with Gammon Bombs rated **Fearless Veteran**.

CURRAHEE!

The men of Toccoa ran up and down Mount Currahee three to four times a week. "Currahee!", which translates from Cherokee as "Stands Alone", became the 506th Parachute Infantry Regiment's battle cry.

Combat and Weapons Platoons led by Toccoa Sergeants may re-roll failed Motivation Tests to rally from being Pinned Down.

CORPORAL DARRELL 'SHIFTY' POWERS

Powers joined Easy Company at Camp Toccoa in 1942. An able hunter, Powers quickly demonstrated his skill with a rifle, making Expert Rifleman and earning a reputation as the company's best shot. Powers jumped into Normandy and Holland and fought alongside his comrades in the Battle of the Bulge.

Like the other men of Easy Company, Powers spent a good portion of time on the perimeter overlooking Foy. Then one day, Powers reported to First Sergeant Lipton that he had spotted a tree that wasn't there the day before. Lipton grabbed a pair of binoculars, but couldn't see what Shifty was talking about. After studying the location for a minute, Lipton noticed gun barrels, and Shifty's 'tree' turned out to be camouflage the Germans were adding to conceal a battery of guns. Lipton radioed in the coordinates of Shifty's 'tree' to the artillery and it was quickly wiped out.

During the battle at Foy, a German sniper had shot three members of Easy Company before he was located and pointed out to Powers. Shifty brought his rifle to bear on the sniper and shot him between the eyes. Many Easy Company veterans reckoned that Powers saved many lives with his deadly skills.

Corporal Darrell 'Shifty' Powers.

Corporal Darrell Powers replaces a Rifle or Rifle/MG team (that is not a Command team) from a Parachute Rifle Platoon (page 16) for +15 points. He is a Warrior Rifle team with Gammon Bombs rated **Fearless Veteran**.

BEST SHOT IN THE COMPANY

'You know, it just doesn't pay to be shootin' at Shifty when he's got a rifle.'

Once per turn, Corporal Powers may re-roll any one failed To Hit roll.

In addition, his rifle has a Firepower rating of 4+, rather than the usual 6.

PARACHUTE RIFLE COMPANY

INFANTRY COMPANY

MOTIVATION AND SKILL

Already veterans of the Normandy and Market Garden campaigns, the paratroopers of the 501st, 502nd, and 506th Parachute Infantry Regiments are well used to fighting while surrounded. A Parachute Rifle Company is rated **Fearless Veteran.**

RELUCTANT	CONSCRIPT
CONFIDENT	TRAINED
FEARLESS	**VETERAN**

You must field one platoon from each box shaded black and may field one platoon from each box shaded grey.
Support platoons can be of any variant type and do not have to match your company ratings.

HEADQUARTERS

- HEADQUARTERS: Parachute Rifle Company HQ 17

COMBAT PLATOONS

- INFANTRY: Parachute Rifle Platoon 17
- INFANTRY: Parachute Rifle Platoon 17
- INFANTRY: Parachute Rifle Platoon 17

WEAPONS PLATOONS

- MACHINE-GUNS: Parachute Machine-gun Platoon 18
- ARTILLERY: Parachute Mortar Platoon 18

SUPPORT PLATOONS

- ARMOUR: Tank Platoon 26; Tank Destroyer Platoon 27
- ARMOUR: Tank Destroyer Platoon 27
- ANTI-TANK: Airborne Anti-tank Platoon 19
- INFANTRY: Parachute Rifle Platoon 17; Glider Rifle Platoon 23; Airborne Engineer Combat Platoon 21; Armored Rifle Platoon 29; Rifle Platoon 29
- RECONNAISSANCE: Cavalry Recon Platoon 28; Airborne Divisional Recon Platoon 28
- ARTILLERY: Parachute Field Artillery Battery 19; Glider Field Artillery Battery 25; Armored Field Artillery Battery 30
- ARTILLERY: Parachute Field Artillery Battery 19; Glider Field Artillery Battery 25; Field Artillery Battery (155mm) 30
- ARTILLERY: Field Artillery Battery (155mm) 30
- ANTI-AIRCRAFT: Anti-aircraft Artillery (Self-propelled) Platoon 31; Airborne Anti-aircraft Artillery Platoon 18
- AIRCRAFT: Air Support 31; Air Observation Post 31

SPECIAL RULES

A Parachute Rifle Company uses all the US special rules on pages 236 to 240 of the rule book in addition to the Master Sergeant special rule.

MASTER SERGEANT

Parachute missions are tricky and the loss of an officer can result in the failure of the mission. Therefore platoon sergeants are briefed on every detail of the mission in the event their officer is separated from the platoon.

Parachute Rifle, Airborne Divisional Recon, and Airborne Engineer Combat Platoons use the German Mission Tactics special rules (see page 242 of the rulebook).

HEADQUARTERS

Parachute Rifle Company HQ

Headquarters

Company HQ	30 points

Options

- Replace Command Carbine teams with Command SMG teams for +15 points per team.
- Add up to three Sniper teams for +50 points per team.

All Carbine or SMG teams in the Company HQ carry Gammon Bombs giving them Tank Assault 3.

The company commander of a 101st Airborne Division Parachute Rifle Company is a vastly experienced veteran of Normandy and Market Garden. His men have complete faith in him and know his leadership will see them through.

COMBAT PLATOONS

Parachute Rifle Platoon

Platoon

HQ Section and Mortar Squad with:

3 Rifle Squads	265 points
2 Rifle Squads	205 points

Options

- Replace Command Rifle/MG team with Command SMG team at no cost.
- Add Bazooka team for +20 points.
- Replace up to one Rifle/MG team in any or all Rifle Squads with a Bazooka team for +5 points per Bazooka team.
- Replace all Rifle/MG teams with Rifle teams for -10 points per Rifle Squad.

All Rifle, Rifle/MG, and SMG teams in a Parachute Rifle Platoon carry Gammon Bombs giving them Tank Assault 3.

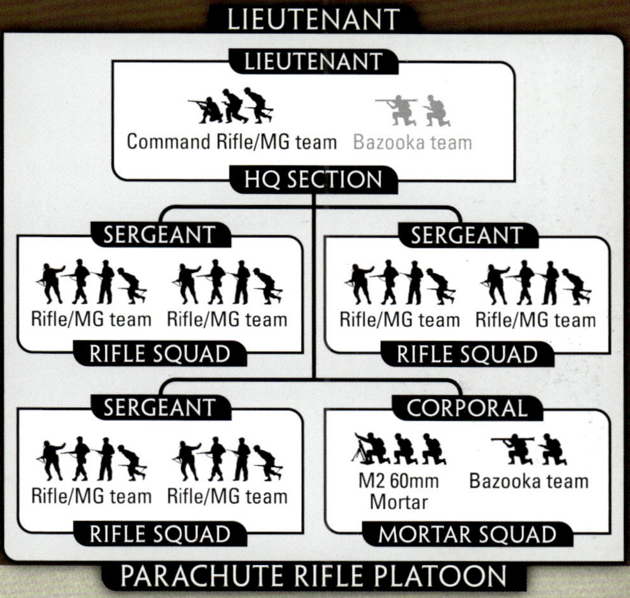

Due to the haste of their departure from Mourmelon camp, there was a shortage of weapons among the men when the 101st Airborne arrived at Bastogne. Additional weapons were taken from retreating 28th Infantry Division soldiers and found discarded along the roads. Eventually, the platoons became fully equipped, but some Parachute Rifle Platoons entered combat without their full complement of weapons.

WEAPONS PLATOONS

Parachute Machine-gun Platoon

Platoon

HQ Section with:

2 MG Sections	170 points
1 MG Sections	100 points

Option

- Add Bazooka teams for +20 points per team.

A Parachute Machine-gun Platoon may make Combat Attachments to Parachute Rifle Platoons.

The additional firepower of the paratroopers' M1919 light machine-gun proved vital during the defence of Bastogne.

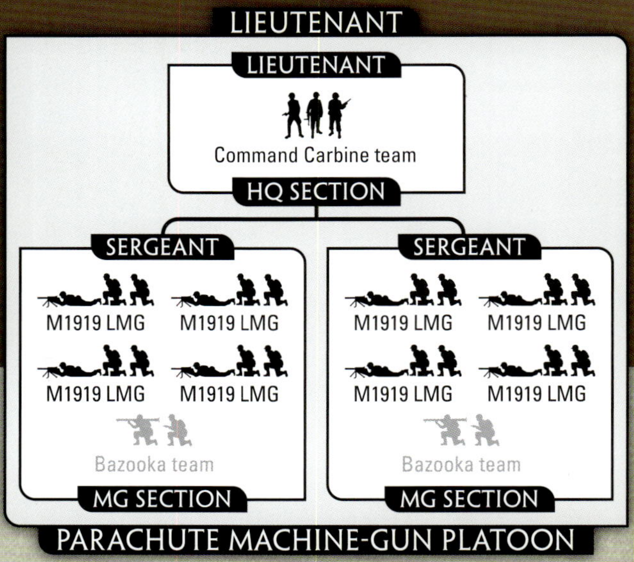

Parachute Mortar Platoon

Platoon

HQ Section with:

2 Mortar Sections	125 points
1 Mortar Sections	65 points

Option

- Add Bazooka teams for +20 points per team.

German troops continually probed the Bastogne perimeter for gaps and weak spots. When the Germans attacked, the paratroopers could quickly call on the mortars to drop a bombardment where it was needed. When one attack was broken up, the mortars could quickly turn to await the next target to present itself.

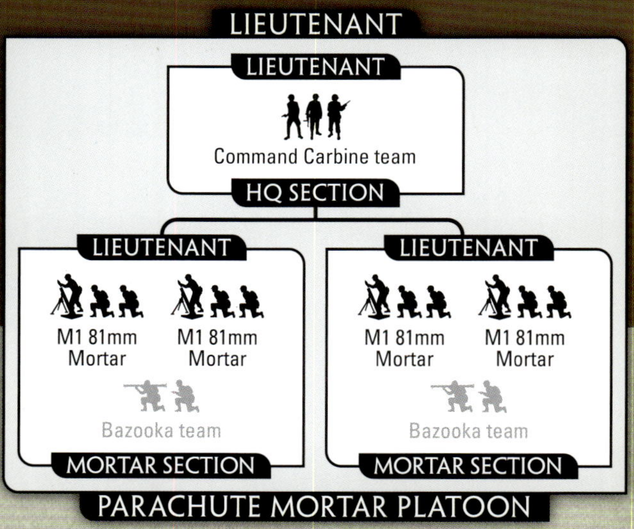

SUPPORT PLATOONS

Airborne Anti-aircraft Artillery Platoon

Platoon

HQ Section with:

2 Automatic Weapon Sections	95 points
1 Automatic Weapon Section	50 points

The airborne anti-aircraft artillery platoon is equipped with the M2 .50cal AA MG on a light tripod. The tall mount and high rate of fire of the gun allows the paratroopers to fill the sky with deadly .50 cal bullets.

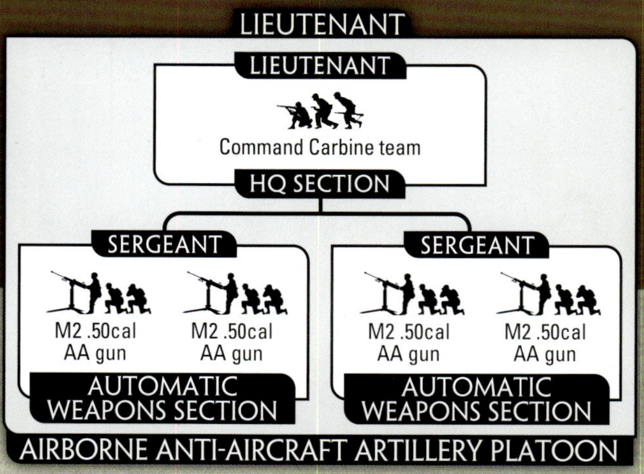

Airborne Anti-tank Platoon

Platoon

HQ Section with:

4 Gun Sections	145 points
3 Gun Sections	110 points
2 Gun Sections	75 points

Options

- Add Jeeps for +5 points for the platoon.
- Add Bazooka teams for +20 points per team.

The M1 57mm anti-tank gun is normally delivered into combat in a glider with its crew, but the 101st Airborne Division arrived in Bastogne by road, with the gun towed by jeeps.

The airborne anti-tank guns played an important role in defending Bastogne. They offered the paratroopers additional protection from enemy tanks. Positioned along the key roads into Bastogne, they could hold off German armoured probes, giving the rest of the airborne forces time to bring more firepower to bear in the sector.

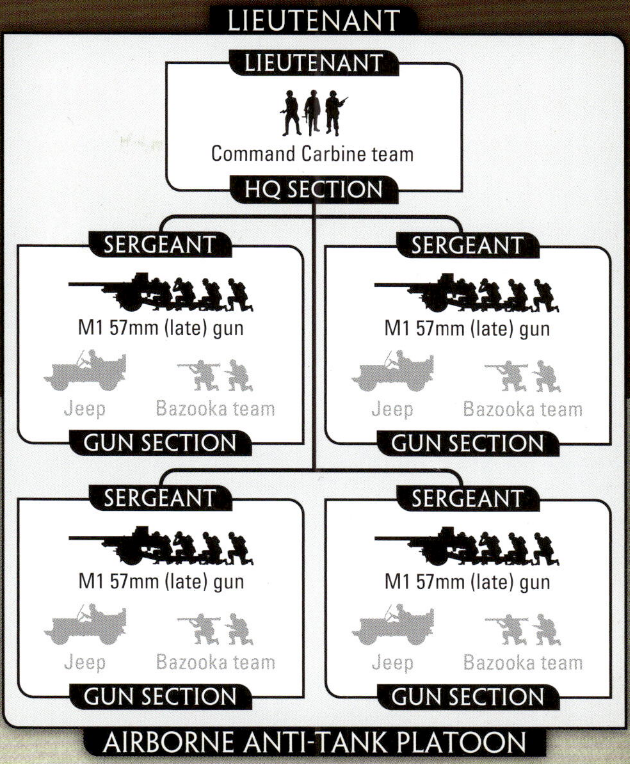

Parachute Field Artillery Battery

Platoon

HQ Section with:

4 M1A1 75mm	135 points
2 M1A1 75mm	75 points

Option

- Add 2½-ton trucks for +5 points for the battery.

The 101st Airborne Division took two types of artillery to Bastogne, the glider artillery battalions armed with M3 105mm howitzers, and the parachute artillery of the 377th and 463rd Parachute Field Artillery Battalions armed with M1A1 75mm pack howitzers. The 463rd Battalion were veterans of the Italian campaign, having spent some time supporting the US/Canadian First Special Service Force, before being temporarily attached to the 101st Airborne Division while waiting for the 17th Airborne Division to arrive from England.

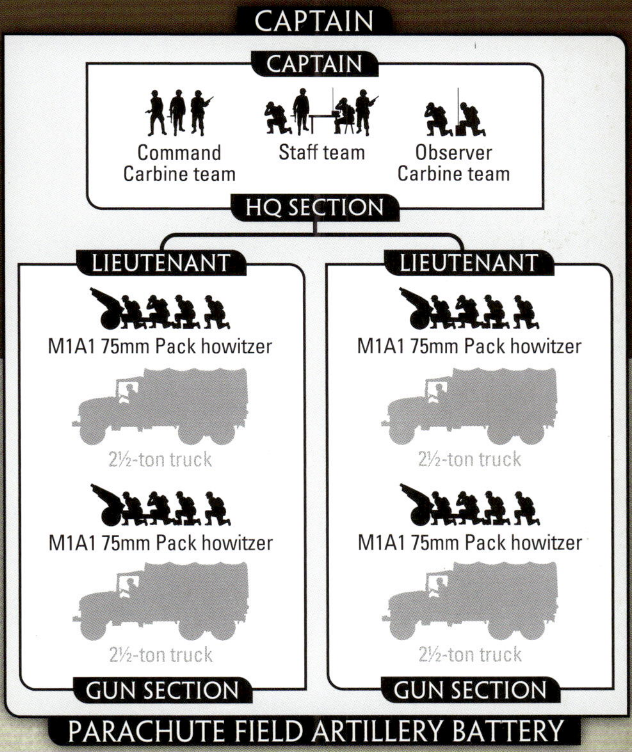

AIRBORNE ENGINEER COMBAT COMPANY
INFANTRY COMPANY

Motivation and Skill

A Glider Airborne Engineer Combat Company is rated **Confident Veteran.**

A Parachute Airborne Engineer Combat Company is rated **Fearless Veteran.**

Glider Troops

RELUCTANT	CONSCRIPT
CONFIDENT	TRAINED
FEARLESS	**VETERAN**

Parachute Troops

RELUCTANT	CONSCRIPT
CONFIDENT	TRAINED
FEARLESS	**VETERAN**

You must field one platoon from each box shaded black and may field one platoon from each box shaded grey.
The Company HQ and Combat Platoons in your force must have the same unit symbol.
Support platoons can be of any variant type and do not have to match your company variant type.

HEADQUARTERS

- HEADQUARTERS: Airborne Engineer Combat Company HQ (21)

COMBAT PLATOONS

- INFANTRY: Airborne Engineer Combat Platoon (21)
- INFANTRY: Airborne Engineer Combat Platoon (21)
- INFANTRY: Airborne Engineer Combat Platoon (21)

SUPPORT PLATOONS

- ARMOUR: Tank Platoon (26); Tank Destroyer Platoon (27)
- ARMOUR: Tank Destroyer Platoon (27)
- INFANTRY: Parachute Rifle Platoon (17); Glider Rifle Platoon (23); Parachute Machine-gun Platoon (18); Glider Machine-gun Platoon (24); Rifle Platoon (29)
- RECONNAISSANCE: Cavalry Recon Platoon (28)
- ARTILLERY: Parachute Field Artillery Battery (19); Glider Field Artillery Battery (25); Armored Field Artillery Battery (30)
- ARTILLERY: Parachute Field Artillery Battery (19); Glider Field Artillery Battery (25); Field Artillery Battery (155mm) (30)
- ARTILLERY: Field Artillery Battery (155mm) (30)
- ANTI-AIRCRAFT: Anti-aircraft Artillery (Self-propelled) Platoon (31); Airborne Anti-aircraft Artillery Platoon (18)
- AIRCRAFT: Air Support (31); Air Observation Post (31)

Special Rules

An Airborne Engineer Combat Company uses all the US special rules on pages 236 to 240 of the rule book in addition to the Master Sergeant special rule.

Master Sergeant

Parachute Rifle and Airborne Engineer Combat Platoons use the German Mission Tactics special rules (see page 242 of the rulebook).

HEADQUARTERS

AIRBORNE ENGINEER COMBAT COMPANY HQ

HEADQUARTERS

Company HQ	15 points	20 points

OPTION

- Replace Command Carbine teams with Command SMG teams for +15 points per team.

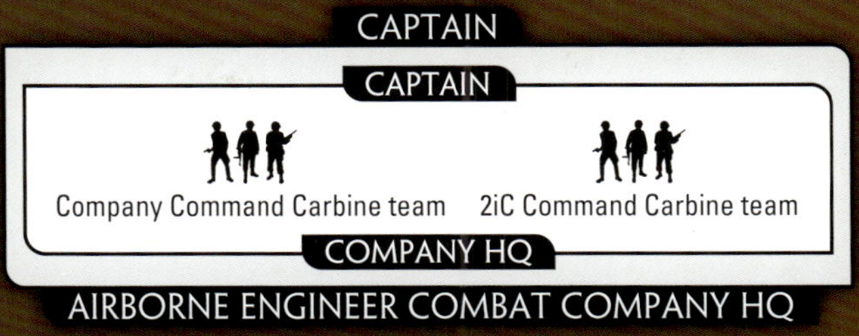

The 326th Airborne Engineer Battalion held the southern approaches around Marvie and were the first to greet the 4th Armored Division as they arrived in Bastogne to relieve the siege.

COMBAT PLATOONS

AIRBORNE ENGINEER COMBAT PLATOON

PLATOON

HQ Section and Weapons Squad with:

2 Operating Squads	170 points	215 points
1 Operating Squads	125 points	155 points
No Operating Squads	80 points	100 points

OPTIONS

- Add Bazooka teams for +20 points per team.
- Add a Pioneer Supply Handcart for +20 points or Pioneer Supply Jeep and trailer for +25 points.

M1919 LMG TEAMS

You may replace any or all Pioneer M1919 LMG teams with Pioneer Rifle teams at the start of the game before deployment.

During Bastogne, the airborne engineers fought as infantry where the firepower of their machine-guns proved its worth. They could also lay mines and were able to knock out a block house to the south of Marvie with explosives.

SUPPLY HANDCART

A Pioneer Supply Handcart operates as a Pioneer Supply Vehicle (see pages 51 and 263 of the rulebook).

GLIDER RIFLE COMPANY

INFANTRY COMPANY

Motivation and Skill

The 327th Glider Infantry Regiment has been through the campaigns in Normandy and Holland and have gained much experience fighting alongside the Paratroopers. A Glider Rifle Company is rated **Confident Veteran.**

RELUCTANT	CONSCRIPT
CONFIDENT	TRAINED
FEARLESS	**VETERAN**

You must field one platoon from each box shaded black and may field one platoon from each box shaded grey.

Support platoons can be of any variant type and do not have to match your company ratings.

HEADQUARTERS

- HEADQUARTERS: Glider Rifle Company HQ — 23

COMBAT PLATOONS

- INFANTRY: Glider Rifle Platoon — 23
- INFANTRY: Glider Rifle Platoon — 23
- INFANTRY: Glider Weapons Platoon — 23

WEAPONS PLATOONS

- MACHINE-GUNS: Glider Machine-gun Platoon — 24
- ARTILLERY: Glider Mortar Platoon — 24
- ANTI-TANK: Glider Anti-tank Platoon — 25

SUPPORT PLATOONS

- ARMOUR: Tank Platoon — 26; Tank Destroyer Platoon — 27
- ARMOUR: Tank Destroyer Platoon — 27
- ANTI-TANK: Airborne Anti-tank Platoon — 19
- INFANTRY: Glider Rifle Platoon — 23; Parachute Rifle Platoon — 17; Airborne Engineer Combat Platoon — 21; Armored Rifle Platoon — 29; Rifle Platoon — 29
- RECONNAISSANCE: Cavalry Recon Platoon — 28
- ARTILLERY: Parachute Field Artillery Battery — 19; Glider Field Artillery Battery — 25; Armored Field Artillery Battery — 30
- ARTILLERY: Parachute Field Artillery Battery — 19; Glider Field Artillery Battery — 25; Field Artillery Battery (155mm) — 30
- ARTILLERY: Field Artillery Battery (155mm) — 30
- ANTI-AIRCRAFT: Anti-aircraft Artillery (Self-propelled) Platoon — 31; Airborne Anti-aircraft Artillery Platoon — 18
- AIRCRAFT: Air Support — 31; Air Observation Post — 31

Special Rules

A Glider Rifle Company uses all the US special rules on pages 236 to 240 of the rule book.

HEADQUARTERS

Glider Rifle Company HQ

Headquarters

Company HQ	15 points

Options

- Replace Command Carbine teams with Command SMG teams for +15 points per team.
- Add up to three Sniper teams for +50 points per team.

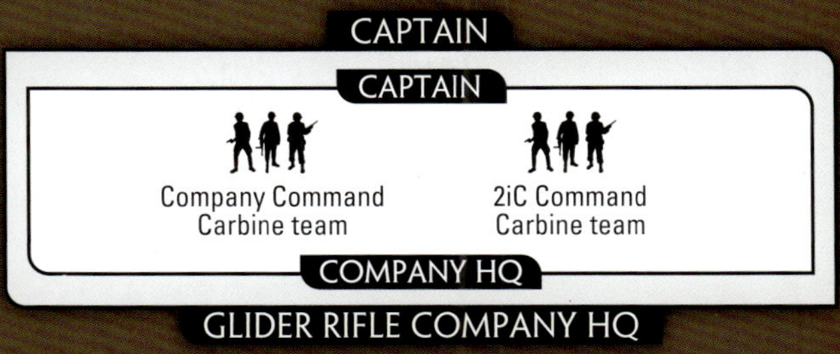

The companies of the 327th Glider Infantry Regiment are organised much like the rifle companies of an infantry division, but with lighter weapons especially suited to glider operations.

COMBAT PLATOONS

Glider Rifle Platoon

Platoon

HQ Section with:

3 Rifle Squads	225 points
2 Rifle Squads	170 points

Options

- Replace Command Rifle team with Command SMG team for +5 points.
- Replace up to one Rifle team in any or all Rifle Squads with a Bazooka team for +5 points per Bazooka team.
- Replace up to one Rifle team in any or all Rifle Squads with an M2 .50 cal MG team at no cost.

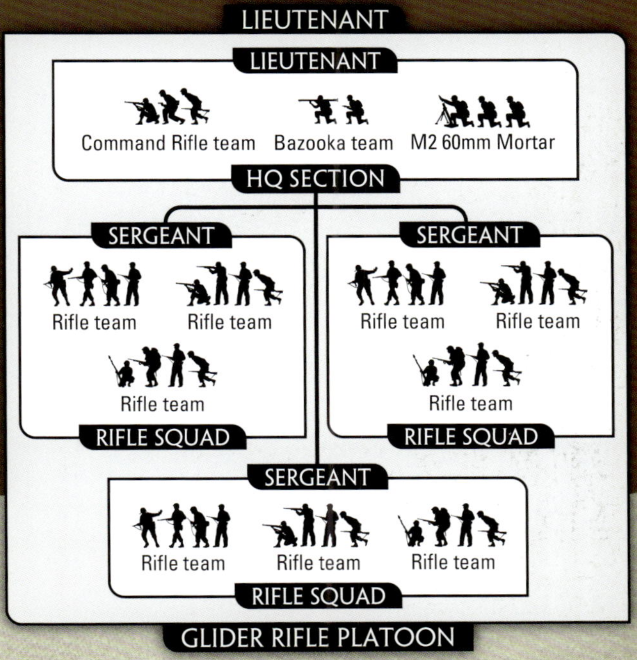

A Glider Rifle Platoon is organised like a standard US Rifle Platoon. They are armed with M1 Garand semi-automatic rifles with additional squad firepower coming from the Browning Automatic Rifle (BAR).

Platoons often have more than one Bazooka, as soldiers have a way of acquiring extra weapons. During the defence of Bastogne the glidermen acquired a number of M2 .50 cal machine-guns from abandoned US half-tracks and pressed them into service as part of their defensive firepower.

Glider Weapons Platoon

Platoon

HQ Section with:

LMG and Mortar Sections	95 points

A Glider Weapons Platoon may make Combat Attachments to Glider Rifle Platoons.

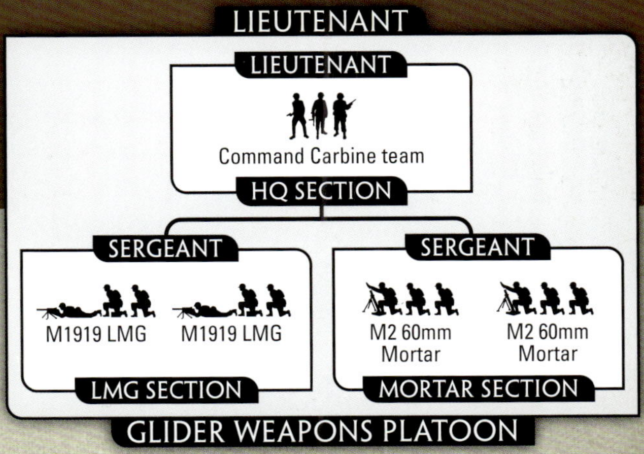

For additional support, the Glider Weapons Platoon provides the glidermen with Browning M1919 light machine-guns and M2 60mm mortars. The light machine-guns are ideal additional firepower in both attack and defence. They are light enough to be moved about the battlefield as required, as are the 60mm mortars. The mortars can provide both supporting bombardments and deadly direct fire.

WEAPONS PLATOONS

Glider Machine-gun Platoon

Platoon

HQ Section with:

4 M1917 HMG	130 points
2 M1917 HMG	70 points

Option

- Add Bazooka teams for +20 points per team.

A Glider Machine-gun Platoon may make Combat Attachments to Glider Rifle Platoons.

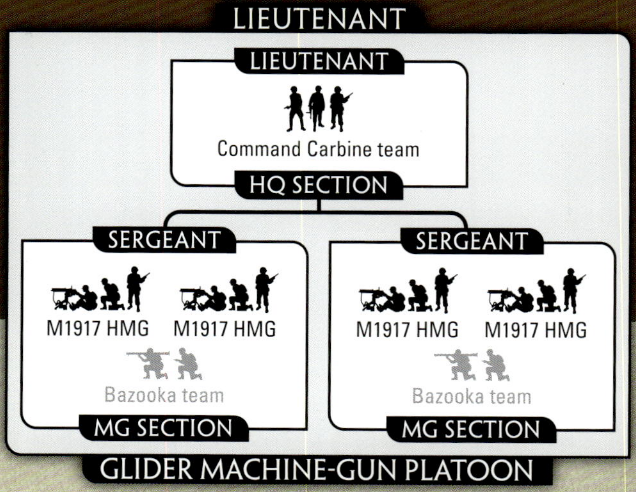

Glider Mortar Platoon

Platoon

HQ Section with:

3 Mortar Sections	155 points
2 Mortar Sections	110 points

Option

- Add Bazooka teams for +20 points per team.

For airborne units the medium mortar takes on extra importance. These compact weapons can be easily loaded on gliders to land with the troops, giving the glidermen readily available artillery as soon as they land.

In Bastogne, while not an airborne operation, they still played an important role in the defence of the town and its important crossroads. The glider mortar crews broke-up a number of attacks on the western flank of Bastogne. Whether using high-explosive rounds or smoke, they provided vital support for the 327th Glider Infantry Regiment's line of defence.

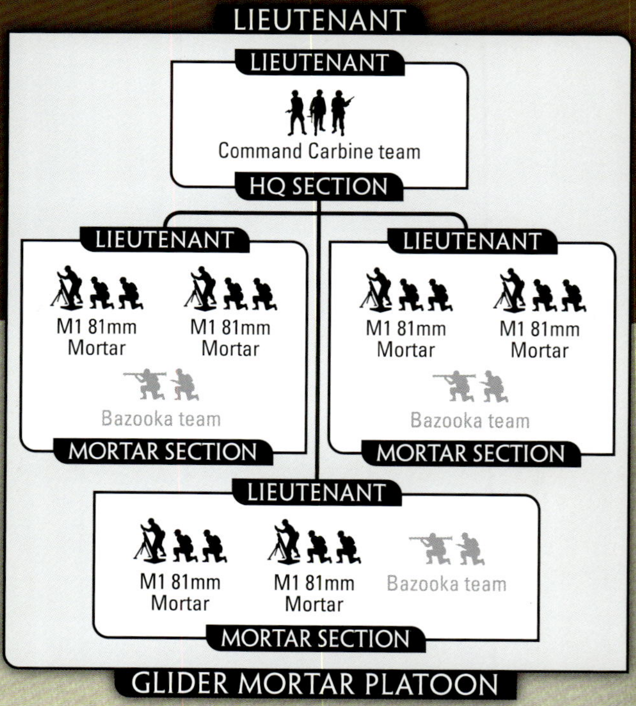

Glider Anti-tank Platoon

Platoon

HQ Section with:

3 M1 57mm (late)	100 points
2 M1 57mm (late)	65 points
3 M3 37mm	70 points
2 M3 37mm	50 points

Options

- Add Jeeps for +5 points for the platoon.
- Add Bazooka teams for +20 points per team.

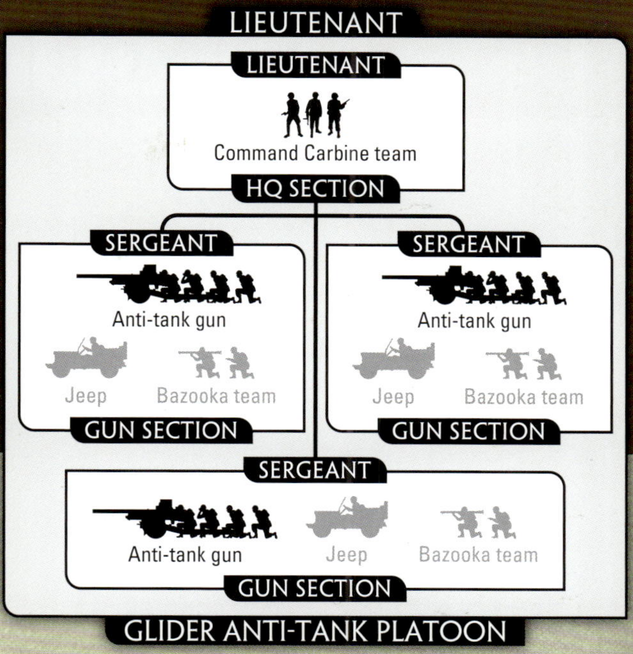

The glider anti-tank units of the 327th Glider Rifle Regiment were initially armed with M3 37mm anti-tank guns, as they would fit in a glider for air landing. By Bastogne, some platoons had been issued with the harder-hitting M1 57mm gun. As well as better anti-tank capability, the 57mm gun has also just been issued a high-explosive round.

Glider Field Artillery Battery

Platoon

HQ Section with:

6 M3 105mm	210 points
3 M3 105mm	115 points

Option

- Add Jeeps for +5 points for the battery.

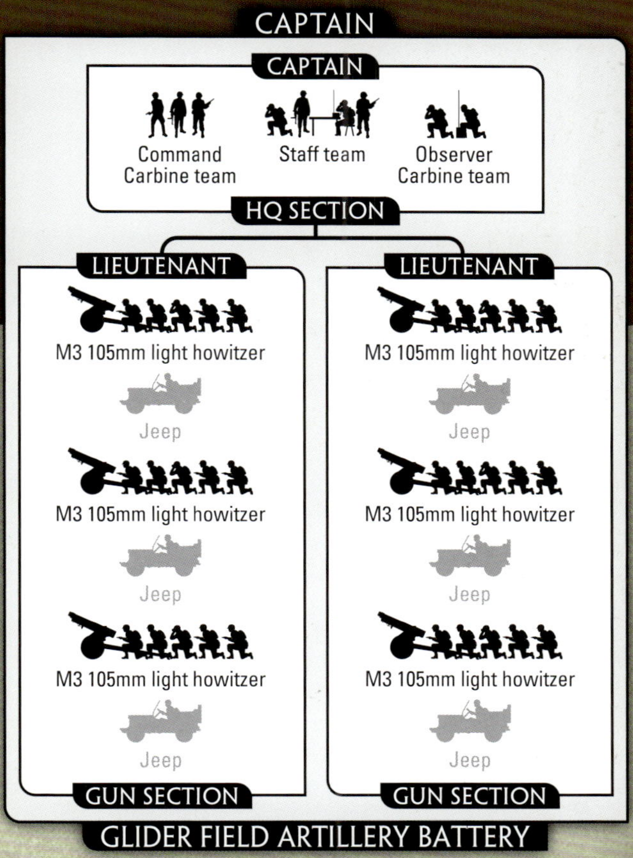

The 321st and 907th Glider Field Artillery Battalions were armed with the M3 105mm light howitzer. The battalions supported the 327th Glider Infantry Regiment, as well as the Parachute Infantry Regiments.

The M3 105mm howitzer offers good punch in a light-weight package. While its bombardment range is not as good as a standard M2A1 field howitzer, but it makes up for that with superior mobility and its ability to be transported by glider.

During the siege of Bastogne, the glider artillerymen ran low of M3 ammunition, so they resorted to using M2A1 ammunition scrounged up by their ammunition and supply men. However, they had to reduce the number of powder charges from seven to three to avoid damaging their howitzers.

US SUPPORT

Motivation and Skill

Fresh armoured troops from the 10th Armored Division's CCB and 9th Armored Division's CCR supported the 101st Airborne Division in Bastogne. The veteran 4th Armored Division arrived to relieve the siege on 26 December.

Veteran Support

RELUCTANT	CONSCRIPT
CONFIDENT	TRAINED
FEARLESS	VETERAN

Trained Support

RELUCTANT	CONSCRIPT
CONFIDENT	TRAINED
FEARLESS	VETERAN

Tank Platoon

Platoon	Veteran	Trained
5 M4 or M4A1 Sherman	415 points	320 points
4 M4 or M4A1 Sherman	335 points	255 points
3 M4 or M4A1 Sherman	250 points	190 points
5 M4A3 (76mm)	690 points	545 points
4 M4A3 (76mm)	555 points	435 points
3 M4A3 (76mm)	415 points	325 points

- Fit any or all tanks with Improvised Armour for +5 points per tank.

Platoon	Veteran	Trained
5 M5A1 Stuart	260 points	200 points
4 M5A1 Stuart	210 points	160 points
3 M5A1 Stuart	155 points	120 points

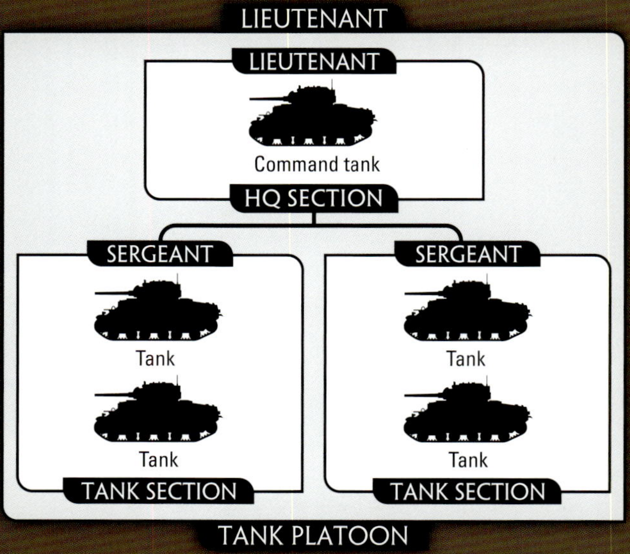

The Sherman tank has served the Army well. Rather than discard the design, the engineers back in Detroit have made them bigger, better, and faster. This new generation of M4s will bring certain victory to the Allied cause.

SHERMAN TANK UPGRADES

You can upgrade your M4 or M4A1 Sherman tanks from the Tank Platoons to newer models. For each tank you wish to upgrade, simply add the points (found to the right) that match the skill rating of the Tank Platoon you have chosen (V or T) to the cost of the platoon.

For example, in your full-strength Veteran (marked V) Tank Platoon (415 points), you would like to upgrade one tank to an M4A3E2 Jumbo, two tanks to M4A3 (76mm), and leave the rest as normal. This adds +60 for the Jumbo upgrade and +55 for each of the 76mm upgrades, for a total cost of 585 points for the platoon.

Replace one M4 or M4A1 tank with:	Veteran	Trained
M4A3E2 Jumbo	+60 points	+50 points

Replace any or all M4 or M4A1 tanks with:	Veteran	Trained
M4A3	+5 points	+5 points
M4A3 (late)	+20 points	+15 points

Replace up to two M4 or M4A1 tanks with:	Veteran	Trained
M4A1 (76mm) (late)	+50 points	+40 points
M4A3 (76mm)	+55 points	+45 points
M4A3E8 Easy Eight	+65 points	+55 points

SHERMAN TANK SPECIAL RULES

In late 1944, the US Army in France started receiving new variants of the M4 Sherman tank. Some of these only differed in their guns and armour, but others changed the way that the tank operated as well. These special rules reflect those changes.

DETROIT'S FINEST

Tanks that use the Detroit's Finest special rule have a Movement Distance of 14"/35cm on Roads or Cross-country Terrain.

DUCKBILLS

At the start of the game a player may elect to fit all of their Sherman tanks (of all variants) with Duckbills. This gives them Wide Tracks (see page 61 of the rulebook), but makes their mobility rating Slow Tank.

M4A3E2 Jumbo and M4A3E8 Easy Eight tanks cannot be fitted with Duckbills and ignore the Duckbills special rule.

SMOOTH RIDE

An M4A3E8 Easy Eight tank does not suffer the +1 penalty to its score To Hit when using the Stabilizers special rule, provided that it did not move more than 6"/15cm during the Movement Step and it did not move in, enter, or move out of Rough Terrain.

JUMBOS LEAD THE WAY

You may allocate hits to an M4A3E2 Jumbo tank as if it had the lowest armour rating, assigning it a hit before the lesser armoured tanks.

This rule does not apply to hits from Artillery Bombardments or hits from Aircraft.

TANK TELEPHONES

If a Tank team with Tank Telephones and an adjacent Infantry team did not move in the Movement Step, and the Infantry team is not Pinned Down, the Infantry team can use the Eyes and Ears rule (see page 195 of the rulebook) to Reveal one Gone to Ground enemy team to that Tank team as if the Infantry team was a Recce team. If other tanks in the platoon fire, they must either have their own Infantry team pointing out the target or continue to treat the target as Gone to Ground.

TANK DESTROYER PLATOON

PLATOON

Security Section and
Tank Destroyer Section with:

	V	T
4 M18 Hellcat GMC	410 points	-
2 M18 Hellcat GMC	230 points	-

OPTION

- Fit any or all Tank Destroyers with Improvised Armour for +5 points per tank.

A Tank Destroyer Platoon uses the US Tank Destroyers special rules on page 238 of the rulebook.

The fast and hard-hitting M18 GMC Hellcats of the 705th Tank Destroyer Battalion made a massive difference in the defence of Bastogne.

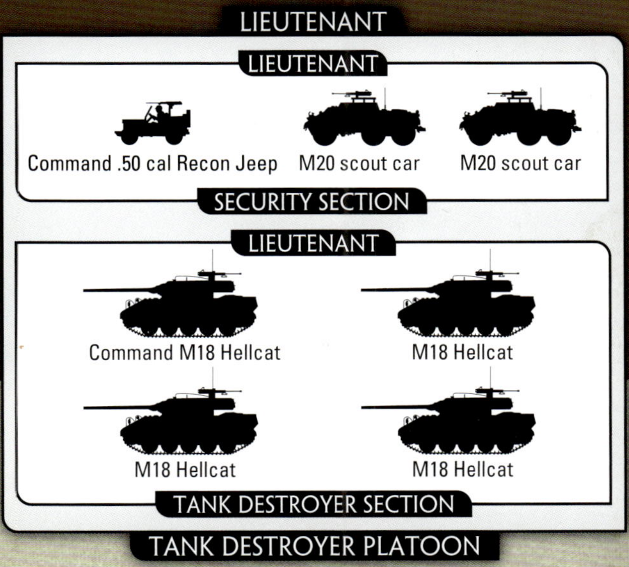

A Tank Destroyer Platoon is a Reconnaissance Platoon.

Cavalry Recon Platoon

Platoon

	V	T
3 Cavalry Recon Patrols	270 points	210 points
2 Cavalry Recon Patrols	180 points	180 points
1 Cavalry Recon Patrol	90 points	70 points

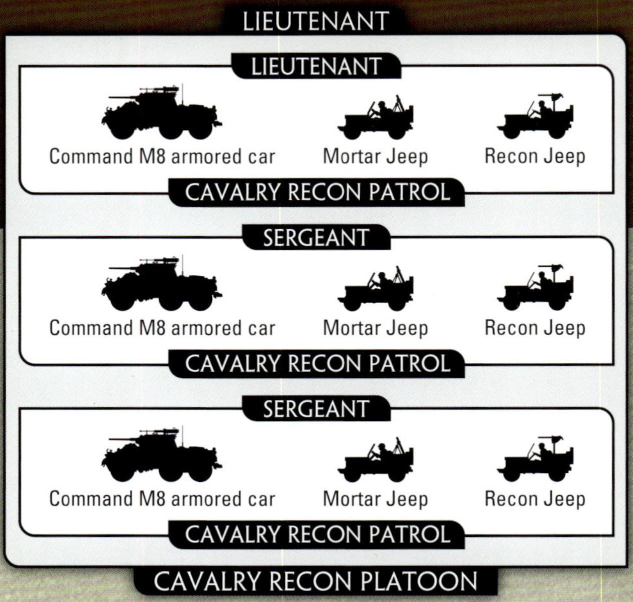

Dismount

Before deployment you may choose to dismount your Cavalry Recon Platoon. If you do this, all of the Cavalry Recon Patrols from the same platoon operate as a single platoon.

If you dismount, all of the platoon's vehicles are permanently removed from the game. Replace all of the vehicles in each Patrol with any two of the following teams for each Patrol:

- *Carbine teams*
- *M1919 LMG teams*
- *up to one M2 .50 cal MG team per Cavalry Recon Patrol.*
- *up to one Bazooka team per Cavalry Recon Patrol.*
- *up to one M2 60mm mortar team per Cavalry Recon Patrol.*

Designate any one of the teams as the Platoon Command team. The platoon remains a Reconnaissance Platoon.

Cavalry Recon Patrols are Reconnaissance Platoons.

Cavalry Recon Patrols from Cavalry Recon Platoons operate as separate platoons, each with their own Command team.

Team O'Hara, CCB, 10th Armored Division had a Cavalry Recon Platoon which fought alongside 501st PIR at Wardin. On 26 December the cavalry recon of the 4th Armored Division arrived in the Bastogne area.

Airborne Divisional Recon Platoon

Platoon

8 .50 cal Recon Jeep	220 points
4 .50 cal Recon Jeep	110 points

Options

- Add a Support Section for +110 points.
- Arm any or all Pioneer Jeeps with an AA MG for +5 points per Jeep.

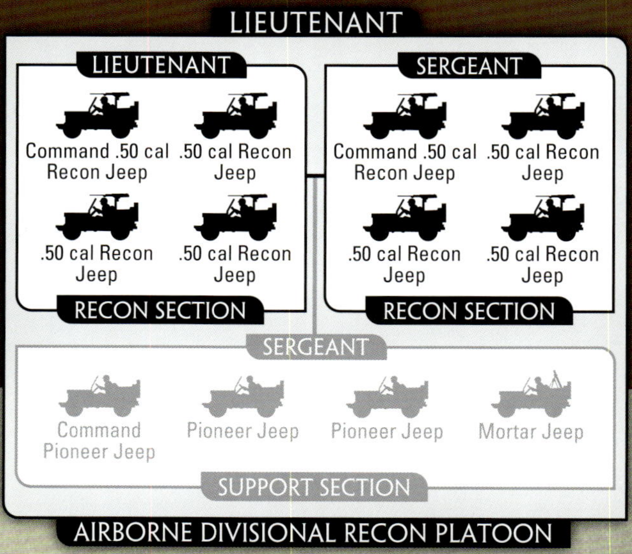

Dismount

Before deployment you may choose to dismount the whole platoon. If you do this, then all of the Recon Sections and the Support Section of an Airborne Divisional Recon Platoon, operate as a single infantry platoon. Designate any one of the teams as the Platoon Command team. The platoon remains a Reconnaissance Platoon.

If you dismount, all of the platoon's vehicles are permanently removed from the game. Dismount the Support Section as normal. Replace all of the vehicles in each Recon Section with any two of the following teams for each Section:

- *Carbine teams*
- *.50 cal MG teams*
- *up to one Bazooka team per Recon Section.*

An Airborne Divisional Recon Platoon is rated **Fearless Veteran.**

FEARLESS | VETERAN

The Recon Sections and the Support Section operate as separate platoons, each with their own command team.

Recon Sections are Reconnaissance Platoons. Pioneer Jeeps are Motorcycle Reconnaissance teams (see pages 196 to 197 of the rulebook) that dismount as Pioneer Carbine teams.

Rifle Platoon

Platoon

HQ Section with:

3 Rifle Squads	155 points
2 Rifle Squads	110 points

Options

- Replace Command Rifle team with a Command SMG team for +5 points.
- Replace up to one team in any or all Rifle Squads with a Bazooka team for +5 points per Bazooka team.

A Rifle Platoon is from the 28th Infantry Division and is rated **Reluctant Veteran.**

RELUCTANT | VETERAN

Many of the men withdrawing from the 28th and other Infantry Divisions were collected up into Team Snafu to help with the defence of Bastogne.

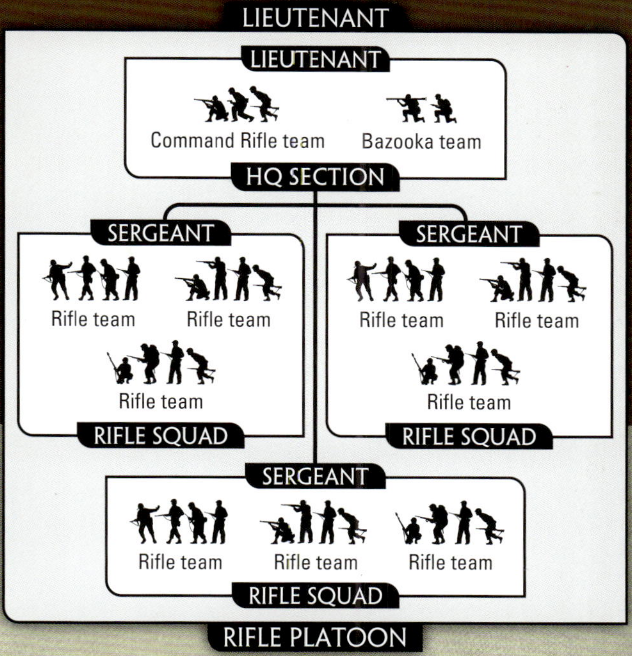

Armored Rifle Platoon

Platoon

HQ Section with Light Machine-gun Squad, 60mm Mortar Squad, and:

	V	T
2 Rifle Squads	295 points	225 points
1 Rifle Squad	235 points	180 points

Option

- Replace Bazooka team in HQ Section with an M3 37mm gun at no cost.

The 54th Armored Infantry Battalion of Team O'Hara, Combat Command B (CCB), 10th Armored Division fought alongside the 3rd Battalion, 501st Parachute Infantry Regiment during the fighting at Wardin and Marvie on 19 to 20 December.

Company B, 20th Armored Infantry Battalion fought as part of Team Desobry alongside the 1st Battalion, 501st Parachute Infantry Regiment.

4th Armored Division's 53rd Armored Infantry Battalion followed behind the 37th Tank Battalion as they finally made contact with the 101st Airborne Division on 26 December, breaking the encirclement and ending the siege.

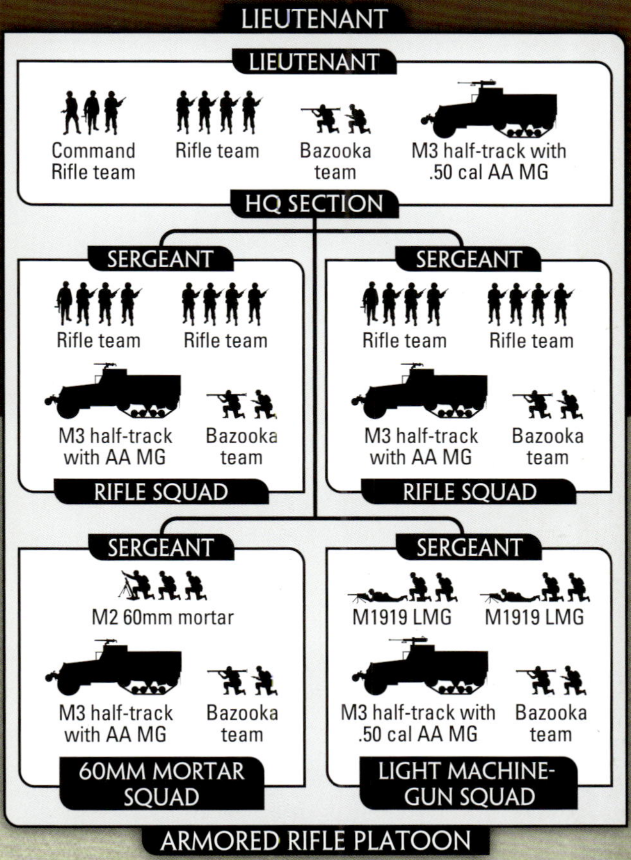

Field Artillery Battery (155mm)

Platoon

HQ Section with:	V	T
2 Gun Sections	-	210 points
1 Gun Section	-	110 points

Options

- Add ¾-ton trucks and M5 high-speed tractors for +5 points for the battery.
- Arm any or all M5 high-speed tractors with a .50 cal AA MG for +5 points per tractor.

The 755th and 969th Field Artillery Battalions, both armed with M1 155mm howitzers, were attached to the 101st Airborne Division during the fighting for Bastogne.

The 969th battalion was unusual because it was made up of African-Americans. It had the remnants of another African-American artillery attached, the 333rd Battalion. The 333rd Battalion was partly overrun on 17 December and withdrew west.

The 155mm howitzers proved a vital part of the defence of Bastogne, providing effective fire support. By the end of the siege, the 969th Battalion had only three operational howitzers left and was almost out of ammunition. For this action, the battalion received the US Presidential Unit Citation and the Belgian Croix d'Guerre with Palm.

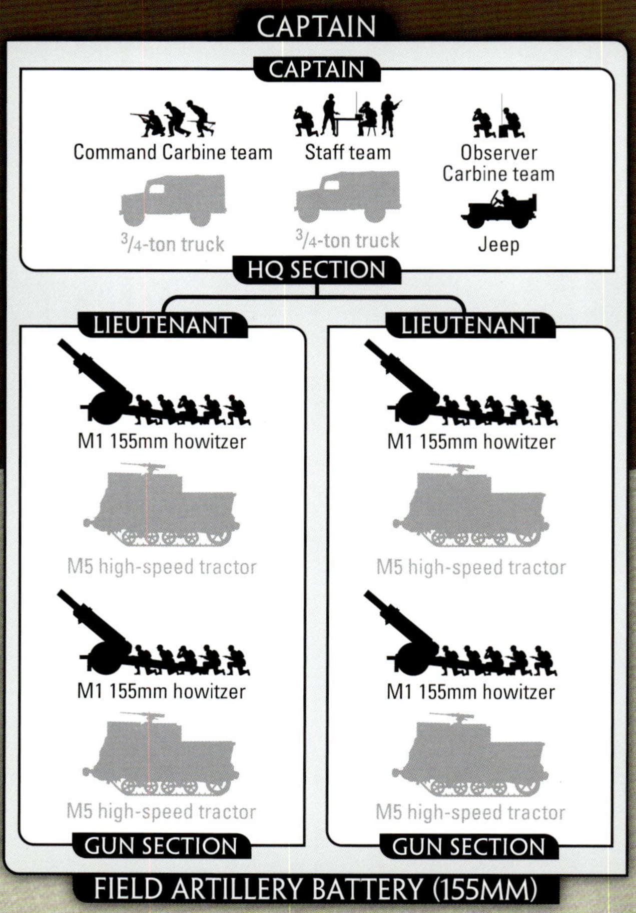

Armored Field Artillery Battery

Platoon

HQ Section with:	V	T
6 M7 Priest GMC	390 points	300 points
3 M7 Priest GMC	220 points	170 points

Options

- Add a Jeep and an M2 half-track with an .50 call AA MG for +5 points for the battery.
- Replace Jeep with an M2 half-track with AA MG for +5 points.

The 420th Armored Field Artillery Battalion of Combat Command B, 10th Armored Division joined the two battalions of towed 155mm howitzers deployed around Bastogne and provided fire support for the glidermen and paratroopers of the 101st Airborne Division during the siege.

After the 4th Armored Division had broken the siege, their artillery was used to support the airborne troops as both divisions went on the counter offensive.

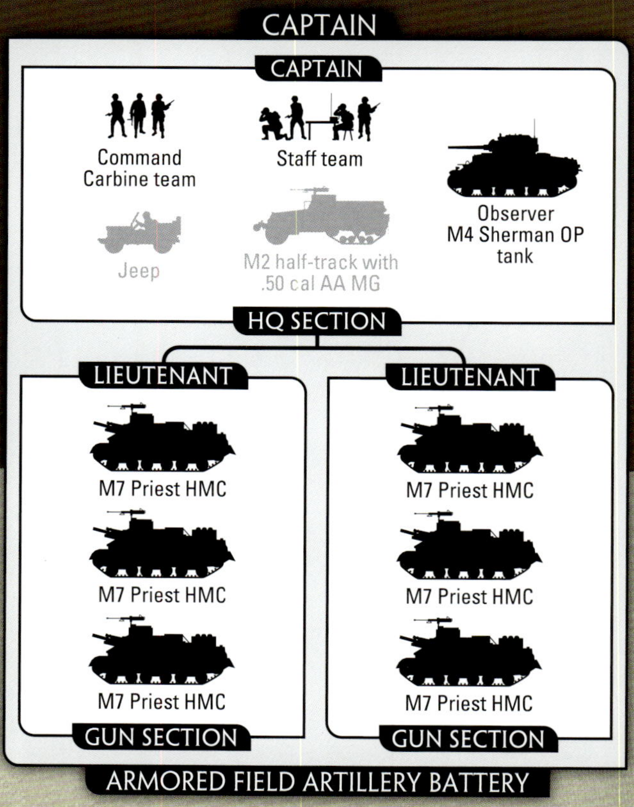

ANTI-AIRCRAFT ARTILLERY (SELF-PROPELLED) PLATOON

PLATOON

Platoon	V	T
2 M16 MGMC (Quad .50 cal) and 2 M15 CGMC (37mm)	200 points	150 points
1 M16 MGMC (Quad .50 cal) and 1 M15 CGMC (37mm)	100 points	75 points

The half-tracks of the self-propelled anti-aircraft artillery (AAA) protect the tanks from both German aircraft and infantry with a hail of machine-gun fire.

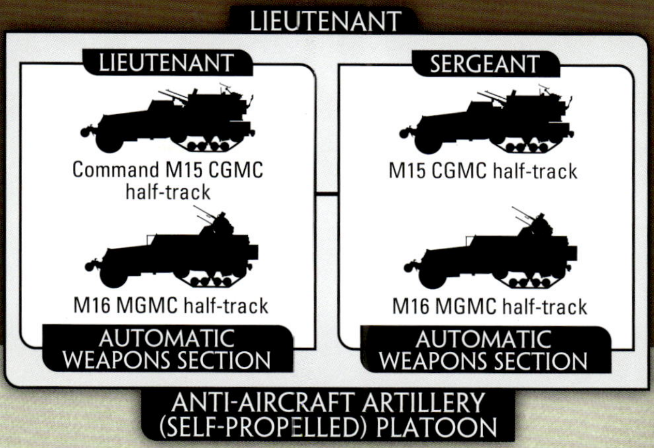

AIR SUPPORT

PRIORITY AIR SUPPORT

P-47 Thunderbolt	190 points

LIMITED AIR SUPPORT

P-47 Thunderbolt	150 points

OPTIONS

- Equip P-47 Thunderbolts with 5.5" HVAR rockets in addition to their normal weapons for +30 points.
- Establish Close Air Support for +25 points.

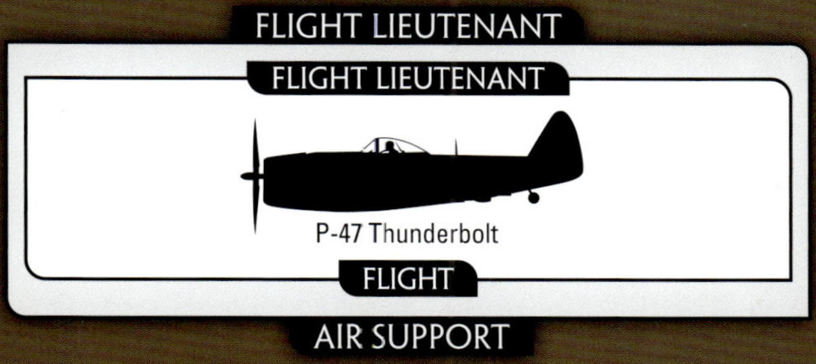

CLOSE AIR SUPPORT

Patton's Third Army enjoyed an unusually close relationship with XIX (19th) Tactical Air Command.

If you establish Close Air Support, you roll two dice on the How Many Aircraft Table and take the best result.

The 406th Fighter Group operating in the skies above Bastogne was the first to use the new 5.5" High Velocity Aircraft Rocket (HVAR), affectionately known as 'Holy Moses' by the crews. This rocket was definitely an improvement over the older, ineffective 4.5" ones used earlier. Each P-47 could carry four of these large rockets, *in addition* to two bombs and its eight .50 cal machine-guns!

AIR OBSERVATION POST

AOP

L4 Grasshopper AOP	40 points

Rare is the day that an L4 isn't seen lingering above the battlefield, directing columns away from danger and calling in devastating artillery bombardments on the enemy.

US ARSENAL

TANK TEAMS

Name *Weapon*	Mobility *Range*	Armour Front *ROF*	Armour Side *Anti-tank*	Armour Top *Firepower*	Equipment and Notes

TANKS

Name *Weapon*	Mobility *Range*	Front *ROF*	Side *Anti-tank*	Top *Firepower*	Equipment and Notes
M5A1 Stuart	Light Tank	4	2	1	Co-ax MG, Hull MG, AA MG.
M6 37mm gun	*24"/60cm*	*2*	*7*	*4+*	*Stabiliser.*
M4 or M4A1 Sherman	Standard Tank	6	4	1	Co-ax MG, Hull MG, .50 cal AA MG, Tank telephone.
M3 75mm gun	*32"/80cm*	*2*	*10*	*3+*	*Smoke, Stabiliser.*
M4A3 Sherman	Standard Tank	6	4	1	Co-ax MG, Hull MG, .50 cal AA MG, Detroit's finest, Tank telephone.
M3 75mm gun	*32"/80cm*	*2*	*10*	*3+*	*Smoke, Stabiliser.*
M4A3 Sherman (late)	Standard Tank	7	4	1	Co-ax MG, Hull MG, .50 cal AA MG, Detroit's finest, Protected ammo, Tank telephone.
M3 75mm gun	*32"/80cm*	*2*	*10*	*3+*	*Smoke, Stabiliser.*
M4A1 (76mm) Sherman (late)	Standard Tank	7	4	1	Co-ax MG, Hull MG, .50 cal AA MG, Protected ammo, Tank telephone.
M1 76mm gun (late)	*32"/80cm*	*2*	*13*	*3+*	*Stabiliser.*
M4A3 (76mm) Sherman	Standard Tank	7	4	1	Co-ax MG, Hull MG, .50 cal AA MG, Detroit's finest, Protected ammo, Tank telephone.
M1 76mm gun (late)	*32"/80cm*	*2*	*13*	*3+*	*Stabiliser.*
M4A3E8 Easy Eight	Standard Tank	7	4	1	Co-ax MG, Hull MG, .50 cal AA MG, Detroit's finest, Protected ammo, Smooth Ride, Tank telephone, Wide tracks.
M1 76mm gun (late)	*32"/80cm*	*2*	*13*	*3+*	*Stabiliser.*
M4A3E2 Jumbo	Slow Tank	12	8	1	Co-ax MG, Hull MG, .50 cal AA MG, Jumbos Lead the Way, Tank telephone.
M3 75mm gun	*32"/80cm*	*2*	*10*	*3+*	*Smoke, Stabiliser.*

TANK DESTROYERS (SELF-PROPELLED)

Name *Weapon*	Mobility *Range*	Front *ROF*	Side *Anti-tank*	Top *Firepower*	Equipment and Notes
M18 Hellcat GMC	Light Tank	2	0	0	.50 cal AA MG, Recce.
M1 76mm gun (late)	*32"/80cm*	*2*	*13*	*3+*	

ARTILLERY (SELF-PROPELLED)

Name *Weapon*	Mobility *Range*	Front *ROF*	Side *Anti-tank*	Top *Firepower*	Equipment and Notes
M7 Priest GMC	Standard Tank	1	0	0	.50 cal AA MG.
M2A1 105mm howitzer	*24"/60cm*	*1*	*9*	*2+*	*Hull mounted, Breakthrough gun, Smoke.*
Firing Bombardments	*72"/180cm*	-	*4*	*4+*	*Smoke bombardment.*
M4 Sherman OP	Standard Tank	6	4	1	Co-ax MG, Hull MG, .50 cal AA MG.
M3 75mm gun	*32"/80cm*	*1*	*10*	*3+*	*Smoke.*

ANTI-AIRCRAFT

Name *Weapon*	Mobility *Range*	Front *ROF*	Side *Anti-tank*	Top *Firepower*	Equipment and Notes
M15 CGMC (37mm)	Wheeled	1	0	0	
M15 37mm combination mount	*24"/60cm*	*4*	*5*	*4+*	*Anti-aircraft.*
M16 MGMC (Quad .50 cal)	Half-tracked	1	0	0	
M45 quad .50 cal gun	*16"/40cm*	*6*	*4*	*5+*	*Anti-aircraft.*

RECONNAISSANCE

Name *Weapon*	Mobility *Range*	Front *ROF*	Side *Anti-tank*	Top *Firepower*	Equipment and Notes
M8 armored car	Wheeled	1	0	0	Co-ax MG, .50 cal AA MG, Recce.
M6 37mm gun	*24"/60cm*	*2*	*7*	*4+*	
M20 Scout car	Jeep	1	0	0	.50 cal AA MG, Recce.
Recon Jeep	Jeep	-	-	-	AA MG, Recce.
.50 Cal Recon Jeep	Jeep	-	-	-	.50 cal AA MG, Recce.
Pioneer Jeep	Jeep	-	-	-	Motorcycle reconnaissance, Dismount as Pioneer Carbine team.
Mortar Jeep	Jeep	-	-	-	Recce.
M2 60mm mortar	*24"/60cm*	*2*	*1*	*3+*	*Hull mounted, Portee, Minimum range 8"/20cm.*
Firing Bombardments	*32"/80cm*	-	*1*	*6*	

Weapon	*Range*	*ROF*	*Anti-tank*	*Firepower*	
VEHICLE MACHINE-GUNS					
Vehicle MG	*16"/40cm*	*3*	*2*	*6*	*ROF 1 if other weapons fire.*
.50 cal Vehicle MG	*16"/40cm*	*3*	*4*	*5+*	*ROF 1 if other weapons fire.*

GUN TEAMS

Weapon	Mobility	Range	ROF	Anti-tank	Firepower	Notes
M2 .50 cal MG	Man-packed	16"/40cm	3	4	5+	
M1919 LMG team	Man-packed	16"/40cm	5	2	6	ROF 2 when pinned down or moving.
M1917 HMG team	Man-packed	24"/60cm	6	2	6	ROF 3 when pinned down or moving.
M2 60mm mortar	Man-packed	24"/60cm	2	1	3+	Minimum range 8"/20cm.
Firing bombardments		32"/80cm	-	1	6	
M1 81mm mortar	Man-packed	24"/60cm	2	2	3+	Smoke, Minimum range 8"/20cm.
Firing bombardments		40"/100cm	-	2	6	Smoke bombardment.
M2 .50 cal AA gun	Heavy	16"/40cm	4	4	5+	Anti-aircraft, Turntable.
M3 37mm gun	Light	24"/60cm	3	7	4+	Gun shield.
M1 57mm gun (late)	Medium	24"/60cm	3	10	4+	Gun shield.
M1A1 75mm pack howitzer	Light	16"/40cm	2	6	3+	Smoke.
Firing bombardments		64"/160cm	-	3	6	Smoke bombardment.
M3 105mm light howitzer	Heavy	16"/40cm	1	7	2+	Breakthrough gun, Smoke.
Firing bombardments		56"/140cm	-	4	4+	Smoke bombardment.
M1 155mm howitzer	Immobile	24"/60cm	1	10	1+	Bunker buster, Gun shield, Smoke.
Firing bombardments		88"/220cm	-	5	2+	Smoke bombardment.

INFANTRY TEAMS

Team	Range	ROF	Anti-tank	Firepower	Notes
Carbine team	8"/20cm	1	1	6	Automatic rifles.
Rifle team	16"/40cm	1	2	6	Automatic rifles.
Rifle/MG team	16"/40cm	2	2	6	
SMG team	4"/10cm	3	1	6	Full ROF when moving.
Bazooka team	8"/20cm	1	10	5+	Tank assault 4.
Staff team	16"/40cm	1	2	6	Moves as a Heavy Gun team.

ADDITIONAL TRAINING AND EQUIPMENT

Pioneer teams and teams equipped with Gammon Bombs are rated as Tank Assault 3.

TRANSPORT TEAMS

Vehicle *Weapon*	Mobility *Range*	Armour Front *ROF*	 Side *Anti-tank*	 Top *Firepower*	Equipment and Notes
TRUCKS					
Jeep or Jeep and trailer	Jeep	-	-	-	Optional Passenger-fired AA MG or .50 cal AA MG.
Dodge ¾-ton, GMC 1½-ton, or GMC 2 ½-ton truck	Wheeled	-	-	-	
M5 high-speed tractor	Standard Tank	-	-	-	Optional Passenger-fired .50 cal AA MG.
Handcart	Wagon	-	-	-	
ARMOURED PERSONNEL CARRIERS					
M2 or M3 half-track	Half-tracked	1	0	0	Optional Passenger-fired AA MG or .50 cal AA MG.

AIRCRAFT

Aircraft	Weapon	To Hit	Anti-tank	Firepower	Notes
P-47 Thunderbolt	MG	2+	6	5+	
	Bombs	4+	5	1+	
	Rockets	3+	6	3+	Optional.

PAINTING AMERICANS

AMERICAN PARACHUTE AND GLIDER INFANTRY

After Normandy the new M1943 olive drab uniform saw widespread issue to airborne divisions. This wasn't just a paratroop uniform, but the beginning of the US Army's push to standardize the combat uniform. All airborne units received the M1943 uniform, including the glider troops, but the paratroopers were quick to modify theirs by adding bigger leg pockets.

For cold weather they were issued Field Drab greatcoats to wear over their uniforms. Some were lucky enough to get black rubber overshoes to protect their feet from the cold and wet.

All colour names and codes given are for the Vallejo range of Flames Of War *paints, available from the online store and* Flames Of War *stockists. More comprehensive painting and modelling guides can also be found on the* ***www.FlamesOfWar.com*** *website.*

AMERICAN VEHICLES

Olive Drab

The painting of US equipment was nothing if not consistent. Almost every tank, truck, and gun was painted the same Olive Drab. Mix a little **Khaki (988)** with **Brown Violet (887)** to add highlights and weathering. The more Khaki you add, the dustier and more faded the vehicle will look. The **US Decal Set (US941)** has markings suitable for most vehicles. For guidance on where to place markings, do an Internet image search.

WINTER VEHICLES

The most obvious method is to paint a whitewash effect over the vehicle's existing colour scheme. With practice, this can look very effective, but it is quite laborious. Build up the colour with several careful coats.

A faster approach is to paint the model white initially, then add scratches and chips in the darker colour, using a fine brush or a sponge-chipping technique. This is good way of simulating a fresh coat of whitewash.

For a streaky, faded whitewash effect, you can start with a whitened version of the basic vehicle colour (pictured is **Green Grey (886)**) and use a combination of the two methods described above.

SNOW BASING

Paint the 'snowy' parts of the base **Sky Grey (989)**.

Selectively drybrush some areas **White (951)**.

Apply a thinned PVA white glue and sprinkle on **Snow flock (GFS027)**.

NORDWIND

German infantry attempt to clear stubborn American defenders from a destroyed village on the Maginot Line.

Flanking M36 Jackson tank destroyers proved the bane of the *Jagdtiger* tank-hunters at Rimling.

Slow, but tough, the Jagdtiger tank-hunter had a gun that would destroy any vehicle it hit.

17. SS panzergrenadiers with supporting *Jagdtiger* tank-hunter approach Rimling as they open Operation *Nordwind*.

A *Flammpanzer 38(t) Hetzer* supports *25. Panzergrenadierdivision* infantry assaulting dug in US infantry.

VOLKSGRENADIER DIVISIONS

The *Volksgrenadier* divisions were formed in the Autumn of 1944 after the destruction of Army Group Centre by the Soviets in the east and of the Fifth Panzer Army in Normandy in the west. The new infantry divisions were designed to economise on personnel and to be primarily focused on defence. These new formations combined the names *Volks* (People's), to inspire a sense of nationalism and fighting spirit, and *grenadier*, to invoke the traditions of the German military.

In an attempt to optimize their fighting power, while economising man-power, each *Volksgrenadierdivision* had three infantry regiments of two battalions (unlike an *Infanteriedivision*, which had three battalions in each regiment). To compensate, the *Volksgrenadier* units were armed with a higher proportion of automatic weapons to create a higher volume of firepower. The *Sturmgewehr 44* (or StG 44) assault rifle was used to arm two of the platoons in each *Volksgrenadierkompanie*, giving them devastating firepower, but only at short ranges. They were also issued with very large numbers of *Panzerfaust* anti-tank weapons, once again short-ranged, but still a powerful asset against Allied tanks. The highly efficient MG42 machine-gun still played an important role with the assault rifle armed *Sturm* (assault) platoons issued with two in the third squad, and three issued to the third *Schützen* (rifle) platoon to provide their main firepower.

The division's supporting arms also economised on personnel. The artillery was organised into larger six-gun batteries (instead of four guns) to save on the number of skilled artillery specialists required in each battalion. This resulted in fewer batteries, but usually the same number of guns or more.

The *Volksgrenadier* divisions retained *Panzerjäger* (anti-tank), *FlaK* (Anti-aircraft), *Pionier* (combat engineer), and *Füsilier* (light infantry and recon) companies or battalions like the old *Infanterie* divisions.

The divisions were organised around a core of veterans from divisions destroyed in the east or west and filled out with new conscripts or excess personnel from the *Luftwaffe* (air force) and *Kriegsmarine* (navy). The mix of recruits, size of the veteran cadre, and the time the new divisions had to form meant the performance of the divisions varied considerably. However, the *Volksgrenadier* divisions played a vital role in the defence of Germany through 1944 and 1945.

12. VOLKSGRENADIERDIVISION

The *12. Volksgrenadierdivision* was rated as one of the best German infantry divisions to fight in the Ardennes offensive. They were assigned the role of punching through the defences on *Rollbahn C* (attack road C), through Büllingen towards Malmedy, clearing the way for *Kampfgruppe Kuhlmann* of *12. SS-Panzerdivision 'Hitler Jugend'*.

The division has its origins in the *12. Infanteriedivision* formed in Mecklenburg. This division fought in Poland, France and Russia. It became surrounded at Minsk in July 1944 and was forced to surrender.

Those remnants of the division that survived the capture of Minsk were used to form *12. Volksgrenadierdivision*, which entered combat at Aachen in October 1944, as the Allies approached the German border.

ARDENNES OFFENSIVE

They were redirected to take part in the Ardennes offensive in December 1944. The division was commanded by *Generalmajor* Gerhard Engel, who had previously been one of Hitler's military aides. The division's own armour was limited, with only six StuG IV assault guns available from the *Panzerjäger Abteilung*, so the *217. Sturmpanzer Abteilung* (which itself only had eight Brummbär assault tanks) was assigned to them for the offensive.

12. Volksgrenadier division's battle began at 0700 hours on 16 December, following a heavy artillery bombardment of the US lines. The Volksgrenadiers slowly advanced, clogging the roads and tracks, while *Kampfgruppe Kuhlmann* waited impatiently behind them ready to make their dash

for the Meuse River. They finally took their first objective, the village of Losheim, at 0900 hours. *27. Füsilierregiment* then pushed down the railway line between Losheim and Losheimergraban toward *Rollbahn C*.

A destroyed bridge over the railway line on the Büllingen Road brought the advance to a halt. However, this only stopped the towed and motorised elements of the division and *Kampfgruppe Kuhlmann*, while the infantry pressed on. Meanwhile, pioneers from *12. Volksgrenadierdivision, 3. Fallschirmjägerdivision* and *12. SS-Panzerdivision* were brought forward to repair the bridge and the SS tanks soon continued their advance.

BUCHHOLTZ STATION

The advance pushed on with *27. Füsilierregiment* on the left and *48. Grenadierregiment* on the right. The Fusiliers ran into stubborn US resistance at Buchholtz railway station. The station was defended by L Company of the III Battalion, 394th Infantry Regiment and one M18 of the 801st Tank Destroyer Battalion, both of the 99th Infantry Division. As the fighting developed, the US commander reinforced the station with a second rifle company. A firefight broke out amongst the wagons in the railway yard, with the German fusiliers taking cover behind them. The Americans moved up their M18 tank destroyer, which promptly went about knocking out the wagons to deny the Germans cover. Concentrated rifle, machine-gun and mortar fire finally forced the Fusiliers to retreat. Another *Füsilier* company supported by two StuG IV assault guns attempted to flank the station, but ran into the US II Battalion, 394th Infantry Regiment, losing a StuG to anti-tank gun fire. Once again, concentrated mortar fire halted the German push.

LOSHEIMERGRABAN

Meanwhile, to the northwest, *48. Grenadierregiment* had been making their way through the woods, slowed by old German minefields and wire left behind by the retreat a few months earlier. They reached Losheimergraban at about 1200 hours and forced back two companies of the US 394th Infantry Regiment. Despite this, by night fall, the US troops still clung to the crossroads.

On 17 December *12. Volksgrenadierdivision* continued to attack with *48. Grenadierregiment*, while *27. Füsilierregiment* swung around the flank from the west, pushing between the I and II Battalions of the US 394th Infantry Regiment. The Americans fell back towards Mürringen and the Losheimergraban crossroads were in German hands.

12. VOLKSGRENADIERDIVISION 16-21 DECEMBER 1944

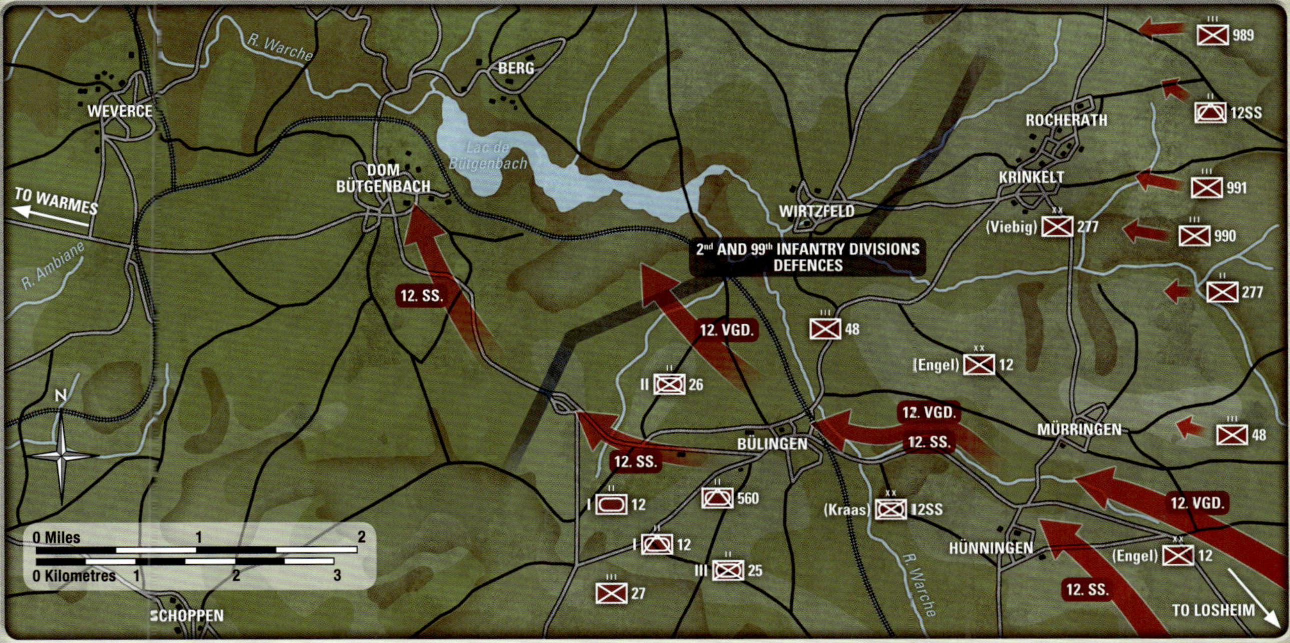

Hünningen

The Volksgrenadiers next thrust was directed towards Hünningen, where the freshly arrived US I Battalion, 26th Infantry Regiment of the 2nd Infantry Division had taken up positions. The *27. Füsilierregiment* led the way once again, followed by the grenadiers and lead elements of *Kampfgruppe Kuhlmann*. Twelve *12. SS Panzer* IV J tanks joined the attack on Hünningen, but the Americans had deployed 3" anti-tank guns. The US guns knocked out four tanks with just their first six shots. The fusiliers bravely attacked in seven separate assaults, but the Americans held firm.

Dom Bütgenbach

The morning of 18 December dawned to find that the Americans had withdrawn during the night to take up new positions at Elsenborn. The *12. Füsilier Bataillon* and five StuG IV assault guns quickly rushed on to Büllingen, which they stormed and captured. They then advanced on Dom Bütgenbach, supported by *Kampfgruppe Krause* (the panzergrenadiers and artillery of *12. SS*).

The US 26th Infantry Regiment was deployed on the high ground around Dom Bütgenbach with the II Battalion, supported by five M4 Shermans (745th Tank Battalion) and four M10s (634th Tank Destroyer Battalion) in concealed positions. The III Battalion was on the II Battalion's left and the I Battalion was at Elsenborn in reserve. The right flank was covered by the 16th Infantry Regiment, 1st Infantry Division who were spread from Dom Bütgenbach to Waimes.

Kampfgruppe Kuhlmann launched an attack on Dom Bütgenbach at 0225 hours on 19 December with armoured panzergrenadiers, but the mud and heavy artillery fire slowed the advance. They attempted again at 1010 hours, but were halted again.

The following day, the fusiliers of *12. Volksgrenadierdivision* resumed the attack, probing through the woods towards Wirtzfeld to the east of Dom Bütgenbach. Meanwhile, the *12. SS* panzergrenadiers attacked to the west of Büllingen road supported by the Jagdpanthers of *560. Schwere Panzerjägerabteilung*. Two Jagdpanther tank-hunters made it into Dom Bütgenbach, but both were knocked out after they drove past the hidden position of a 57mm anti-tank gun. The *kampfgruppe* then retired. Three more attacks that day were halted by US tank destroyers and artillery fire, despite hefty reinforcement of the panzergrenadiers' attacks with heavy *Volks-Artillerie* fire.

One final attempt was made on 21 December to take Dom Bütgenbach, reinforced by the final arrival of *217. Sturmpanzerabteilung*. Initially the panzergrenadiers, tanks and assault guns made good progress, but the ever watchful US tank destroyers and artillery saved the day once again for the Americans. That night *12. SS-Panzerdivision* troops withdrew from the line, leaving the *12. Volksgrenadierdivision* to hold. The advance had been halted at Dom Bütgenbach.

After the Ardennes, *12. Volksgrenadierdivision* remained fighting in the west, taking part in the fighting for the Roer River line against the US Ninth Army. It ended the war just a remnant of the unit it had been, holding out to the last in the Ruhr Pocket in mid-April 1945.

277. Volksgrenadierdivision

277. Volksgrenadierdivision was created from the remnants of *277. Infanteriedivision* during September and October 1944 after the original division had escaped through the Falaise Gap in August. It was sent to Hungary for training and reformation with a core of about 2500 veterans. Most of the replacement personnel did not arrive until early October. These were mostly made up of the Croatians of *574. Volksgrenadierdivision*.

The division relocated to the West on 30 October, but progress was slow. They finally replaced *347. Infanteriedivision* on 10 November while the latter was relocated to Alsace. They were initially deployed on the fortified West Wall between Wallerscheid and Losheim, but their front expanded as other units were drawn off to take part in the battles for the Huertgen Forest to the north.

The German command harboured doubts about the morale of some troops, especially *Volksdeutsche* (foreign ethnic Germans), Alsatians (who were assumed to hold French sympathies), and soldiers from the Vienna area currently under threat from the Soviets.

Ardennes Offensive

Training for an offensive began in early December, with emphasis on forest and night fighting. The commander of the division, *Oberst* Wilheim Viebig, was also informed of the coming offensive and the role of his division. They were to punch through the defensive line of the Americans and allow the following panzer troops to push through. They were to take the villages of Rocherath and Krinkelt (the 'twin villages' as they became known in US accounts) and clear the Hollerath-Losheim road to Elsenborn for *Kampfgruppe Müller* of *12. SS-Panzerdivision 'Hitler Jugend'*. On their right would be *12. Volksgrenadierdivision*. On their left was *3. Fallschirmjägerdivision*.

Rocherath and Krinkelt

On 16 December in the early morning darkness, after a brief artillery barrage, the division's *989. Volksgrenadierregiment* and *990. Volksgrenadierregiment* penetrated the forward lines of the US 99th Infantry Division. The American division had deployed in depth in the forest and *277. Füsilierkompanie* on the left, attacking on either side of the road to Krinkelt, soon ran into stiff resistance. The *990. Volksgrenadierregiment* to the north, was stopped before even reaching the edge of the wood. Despite reinforcements from *991. Volksgrenadierregiment,* the attack was unable to resume its advance.

The *989. Volksgrenadierregiment,* encountering increasing American resistance, had fought its way through the woods to the Jans Stream nearing the Rocherath-Wahlerscheid road. Communication had also been lost with *277. Füsilierkompanie* who, having not received any support during their advance, had taken up defensive positions on a crossroads 1.2 miles (2km) south of Udenbreth.

The Americans gave some ground against *989. Volksgrenadierregiment* on the northern flank, but generally they resisted strongly throughout the day. It wasn't until the next day, after the division received reinforcements from a panzergrenadier regiment from *Kampfgruppe Müller*, that they were able to push all the way to the western edge of the wood facing Rocherath. With the support of assault guns from *Kampfgruppe Müller* and their own Hetzer assault guns, the Volksgrenadiers were able to gradually occupy Rocherath and Krinkelt. They then pushed to the northwest, occupying the woods as the Americans withdrew. All three regiments suffered heavy casualties during the two days of fighting, especially amongst the battalion and company commanders. The division took a secondary role as *3. Panzergrenadierdivision* arrived to continue the assault on the Elsenborn Ridge.

By 20 December *277. Volksgrenadierdivision* had occupied the Kalterherberger Forest and had anchored its right flank on Hill 604, where they took up defensive positions.

After the Battle of the Bulge, the division fought a retreat eastwards and finally surrendered in the Ruhr Pocket in April 1945.

277. VOLKSGRENADIERDIVISION 16-20 DECEMBER 1944

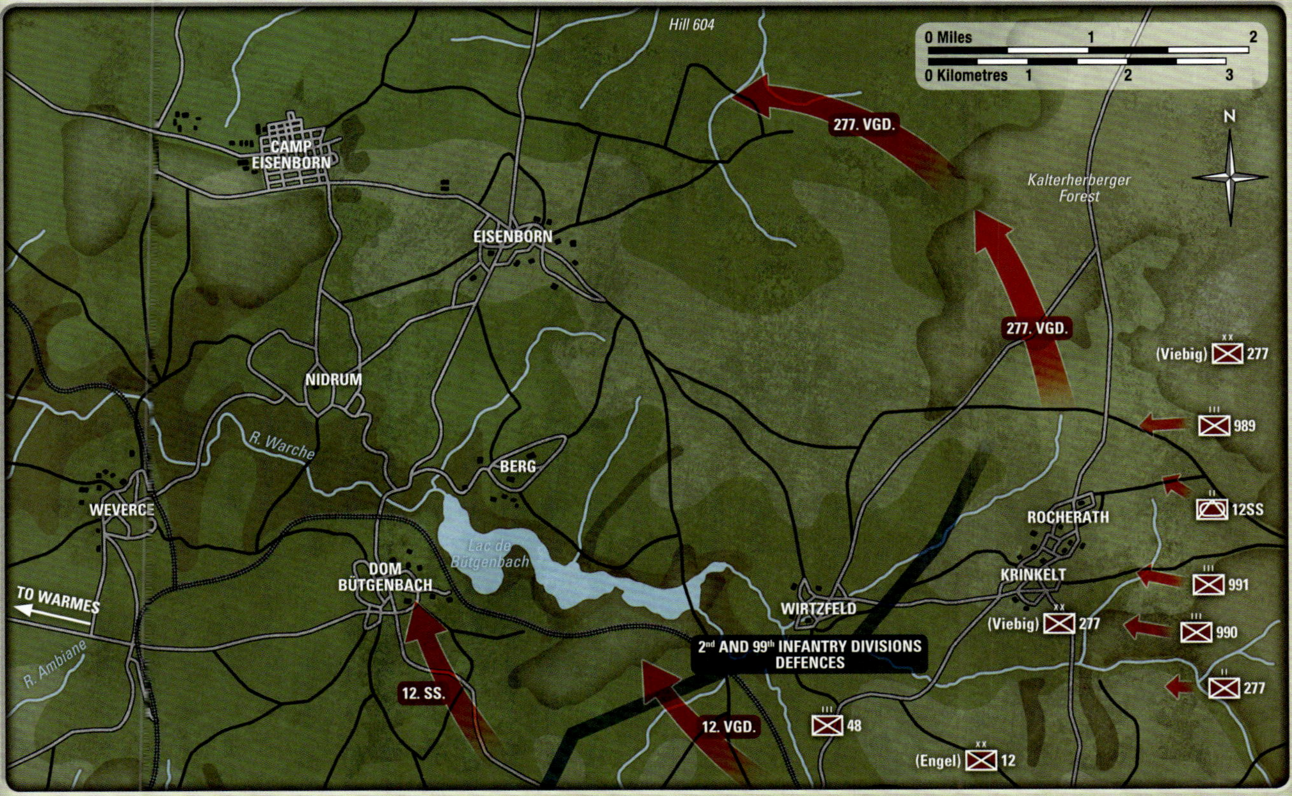

26. VOLKSGRENADIERDIVISION

26. Volksgrenadierdivision formed in September 1944 from remnants of the veteran *26. Infanteriedivision* and the partially formed *582. Volksgrenadierdivision*. The division was filled out with *Luftwaffe* and *Kriegsmarine* replacements.

Ardennes Offensive

The task set for *26. Volksgrenadierdivision* for the Ardennes offensive was tough. They had to cross the Our River, fight their way up its steep western bank, and break through the US lines around Bastogne. In their path lay hundreds of felled trees, copious barbed wire entanglements and large areas of mines, all left behind by retreating German troops earlier in the year. They had to penetrate all of these obstacles and punch through the American lines, thus clearing the way for *2. Panzerdivision* and *Panzer Lehr Division* to push through towards their objectives further behind the American lines.

Under the command of *Oberst* Heinz Kokott, they attacked the US 110th Infantry Regiment of the 28th Infantry Division on 16 December. The American riflemen had deployed along 'Skyline Drive', a ridge running behind the Our River between St Vith and Diekirch. The division quickly pushed through to the villages of Marnach, Hosingen and Holzthum.

Hosingen and Drauffelt

At Hosingen they met III Battalion of the 110th Infantry Regiment, whose K Company, as well as B Company of the 103rd Engineer Combat Battalion, held the village. The village held access to the good road leading west towards Bastogne and over the Clerf River. *77. Volksgrenadierregiment* led the assault towards Hosingen, while *39. Füsilierregiment* pushed along the narrower road to the south towards Holzthum.

On reaching Hosingen I Battalion, *77. Volksgrenadierregiment* was hit with heavy fire and swung around to the north to continue on to Drauffelt. The following II Battalion immediately assaulted Hosingen, but the two American companies were well dug-in and the assault devolved into a series of close range fights, with many hand-to-hand clashes, lasting throughout the day.

I Battalion, *77. Volksgrenadierregiment* continued on to Drauffelt, coming under artillery fire from Buchholz, forcing them to swing north again. They finally captured the bridge over the Clerf River in the early hours of 17 December.

Oberst Kokott also committed *78. Volksgrenadierregiment*. Its II Battalion seized the crossing at Wilwerwiltz, while I Battalion joined the assault on Hosingen with some of the division's Hetzer tank-hunters. The defenders finally surrendered on 17 December. Similar determined defence was encountered at Holzthum and Consthum, both also falling on 17 December.

26. VOLKSGRENADIERDIVISION 16-17 DECEMBER 1944

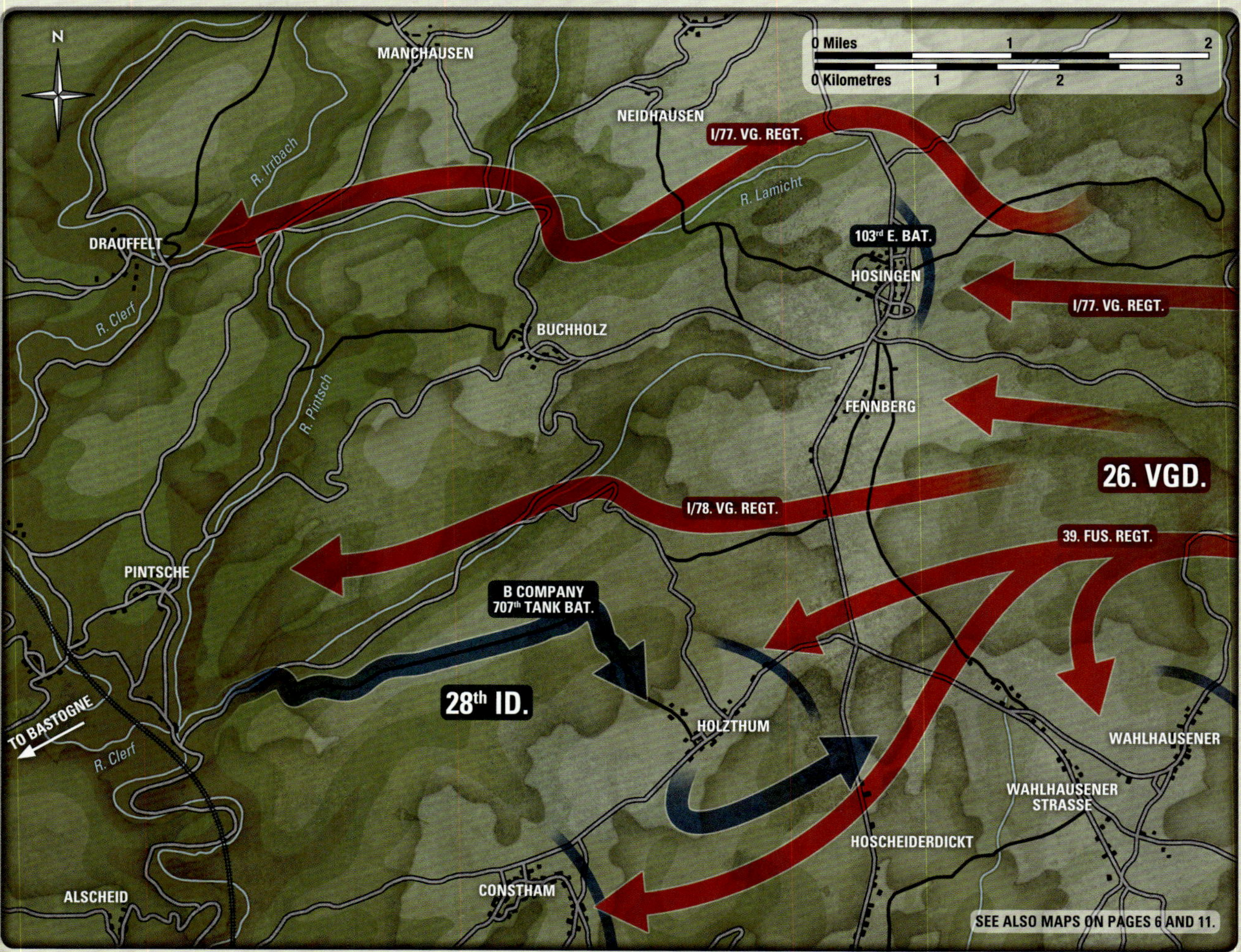

SEE ALSO MAPS ON PAGES 6 AND 11.

Bastogne

The division reached the outlying Bastogne defences at Neffe and Bizory by 19 December. This allowed the two panzer divisions following the *Volksgranadierdivision* to bypass Bastogne to the east and west. However, by this time the American 101st Airborne Division had arrived to take over the defence of Bastogne from the shattered 28th Infantry Division. These elite airborne troops would prove tough to shift.

Between 19 and 22 December *26. Volksgrenadierdivision* made several attempts to breach the American perimeter to no avail. However, they did have the Americans surrounded. The Germans offered the Americans a chance to surrender on 22 December, only to receive the terse reply of 'Nuts!'

The division spent January attempting to break into the paratrooper stronghold, before being forced to fall back when faced with Patton's counter-offensive in February.

The division's lines were finally broken when the Americans crossed the Pruem River. However, the division fought on until they had just a few hundred men left. With their escape route cut by American armoured troops, they finally surrendered just west of the Rhine River in March 1945.

Volksgrenadier Weapons

The key weapons in the Volksgrenadier infantryman's arsenal were the StG44 assault rifle, MG42 light machine-gun, and the *Panzerfaust* anti-tank projectile launcher.

Panzerfaust Anti-tank Projectile Launcher

The *Panzerfaust* worked on the same principles as the recoilless rockets used by the *Fallschirmjäger*. It was actually a rocket, the projectile was launched by a black powder explosive charge. It was a light and disposable one-shot weapon.

It worked by launching a fin stabilised wooden shaft that was tipped with a shaped charge warhead. Like all HEAT (high-explosive anti-tank) rounds it worked by exploding on contact with a vehicle's armour and forcing a pressurised jet of gases and metal particles through the armour plate.

The third variant of the *Panzerfaust*, the *Panzerfaust 60*, was introduced in September 1944. This eventually became the most commonly produced *Panzerfaust*. The 60 denoted its effective range of 60 meters. The *Panzerfaust 60* was issued in large numbers to the newly formed *Volksgrenadier* divisions and was available to these infantrymen in much larger numbers than it had been to other units before them. The *Panzerfaust* gave the Volksgrenadiers a simple and effective anti-tank weapon that could be used by anyone, no matter their level of training, and it could penetrate up to 200mm of armour.

StG44 Assault Rifle

The StG44 assault rifle (*Sturmgewehr 44*, Assault Rifle 44) is an amazingly innovative weapon that gave birth to a whole generation of modern infantry assault rifles used around the world today. It combined the ease of use of a carbine with the firepower of a submachine-gun. The user could select one of two firing modes: semi-automatic, where one squeeze of the trigger gave the grenadier one shot, or fully-automatic, where the weapon fired as long as the grenadier held down the trigger or the weapon run out of ammunition. German manuals recommended it only be used on fully-automatic in emergencies, mainly to ensure that the soldier didn't waste ammunition unnecessarily. It had a lighter round than the standard K98 rifle, but German studies had already established that most combat took place at less than 300 meters and the power of the shorter 7.92mm round was more than enough at these ranges.

MG42 Light Machine-gun

Often referred to as the best machine-gun of the war, the MG42 had an amazing rate of fire of 1200 rounds per minute. The German field manual recommended only firing 300 to 350 rounds per minute to reduce overheating and barrel wear. It had a quick-change barrel so a crew could replace the barrel if they needed to keep firing for a long period. Its high rate of fire gave it a unique ripping sound that was easily recognised. Its sound and reputation soon gave it a terror of its own, with Americans naming it 'Hitler's buzz saw'.

MAJOR SIEGFRIED MOLDENHAUER

Major Siegfried Moldenhauer commanded I Battalion, *48. Volksgrenadierregiment, 12. Volksgrenadierdivision* during the Ardennes offensive. He was a vastly experienced Eastern Front veteran who had already received an impressive number of awards through his bravery, leadership and skill.

Moldenhauer was born in East Prussia in 1915 and joined the Army in 1930. In May 1940 he led a platoon of the *48. Grenadier Regiment, 12. Infanteriedivision* during the battle for France where he received the Iron Cross 2nd Class and was made *Oberfeldwebel* (Master Sergeant).

He again distinguished himself in June and July 1941 taking part in the invasion of the Soviet Union. By August his bravery earned him the Iron Cross 1st Class and the Infantry Assault Badge in Silver. After fighting in the winter battles of 1941/42 he was promoted to *Leutnant* (Lieutenant). His division was encircled at Demjansk, where he took command of his 5th Company and then the 7th Company. The ring was broken on 28 April 1942 and Moldenhauer and his men were able to escape. He received the German Cross in Gold for his leadership.

After promotion to *Hauptmann* (Captain) in March 1943 he was back in action with his regiment at Newel, then Vitebsk. In September 1943 he received the newly-created Close Combat Clasp in Bronze, having already spent over 20 days in close combat. By October he had racked up 30 days and was awarded the Close Combat Clasp in Silver. Each of those days involved some form of combat, either assaults against enemy positions, hand-to-hand fighting, leading patrols, or leading critical counterattacks.

In January 1944 he was awarded the Knight's Cross and made acting commander of II Battalion. It wasn't long before he took command of I Battalion after its commander was killed.

Moldenhauer's division was heavily involved in defensive fighting, especially after the Soviets launched Operation Bagration. *12. Infanteriedivision* became encircled around Mogilev. Leading a rearguard *kampfgruppe*, Moldenhauer extracted his men through a narrow gap in the encirclement. His *kampfgruppe* (a weak battalion in strength) joined other remnants of the division under Major Wilhelm Osterhold who ordered him to established a five kilometre long line to hold the Soviets at bay, while the rest of the division escaped and the engineers prepared bridges across the Dnepr River for demolition. With no heavy weapons to speak of, his *kampfgruppe* held off strong Soviet rifle forces for 24 hours, during which time they destroyed ten tanks, before units of *337. Infanteriedivision* arrived to reinforce their position. Despite twice being wounded he kept fighting and achieved his 50th day of close combat.

In July the survivors of *Kampfgruppe Osterhold* had withdrawn to Poland where they continued to fight against the Soviet advance. Moldenhauer was promoted to Major on 1 July and was awarded the Close Combat Clasp in Gold in October 1944 (the first in his division to achieve this).

After rebuilding and being renamed *12. Volksgrenadierdivision* his division took part in the Ardennes offensive in December. After the initial successful breakthrough of the American lines, he was wounded for the fifth time. He spent the rest of the war in the officer reserve.

CHARACTERISTICS

Major Siegfried Moldenhauer is a Higher Command SMG team rated **Confident Veteran**. Moldenhauer can be added to a Volksgrenadierkompanie for +55 points (page 46).

INFANTRY ASSAULT BADGE

Moldenhauer is never far from the action and is not afraid to mix it up in the thick of the fighting.

Moldenhauer hits on a 2+ in Assaults.

CLOSE COMBAT CLASP IN GOLD

Even before his division began their attack on 16 December 1944, Moldenhauer had spent over 70 days in close combat and was an expert at holding his men together in the face of the enemy.

Any platoon joined by Moldenhauer entirely made up of Infantry teams, with or without Transport teams or Combat Attached Gun teams, that is not Pinned Down at the start of the Assault Step, that becomes Pinned Down in the Assault Step, can take a Motivation Test.

- *If the test is passed, remove the Pinned Down marker.*
- *Otherwise, the platoon remains Pinned Down.*

If the platoon receives five hits from Defensive Fire, they still fall back and the assault is over.

FIELDING VOLKSGRENADIERS

You can use the Volksgrenadierkompanie on page 46 to represent many of the other *Volksgrenadier* divisions fighting all along the Western Front, including the Ardennes and *Nordwind* offensives.

Some divisions retained a strong core of experienced veterans, as well as much of their experience and confidence, like *12. Volksgrenadierdivision* and *26. Volksgrenadierdivision* and use the *12. Volksgrenadierdivision* Motivation and Skill option in the Volksgrenadierkompanie. You can also represent other experienced divisions like *36. Volksgrenadierdivision, 62. Volksgrenadier-division,* and *352. Volksgrenadierdivision,* to name a few.

Many other divisions, like *277. Volksgrenadierdivision,* were formed around a much smaller core of veterans and relied on a higher proportion of new recruits and personnel transferred from the *Luftwaffe* (Air Force) and *Kreigsmarine* (Navy). They tended to have less experience and lacked confidence in battle. Use the *277. Volksgrenadierdivision* Motivation and Skill option in the Volksgrenadierkompanie. As well as *277. Volksgrenadier-division,* you can also represent other less experienced divisions like *9. Volksgrenadierdivision, 18. Volksgrenadierdivision,* and *47. Volksgrenadierdivision,* for example.

VOLKSGRENADIER SPECIAL RULES

A Volksgrenadierkompanie uses all the German special rules on pages 241 to 245 and the following additional special rule.

STOSSTAKTIK

A number of *Volksgrenadier* divisions infiltrated infantry units through the US lines at the start of their operations in the Ardennes!

If you field a Volksgrenadier Scout Platoon your Volksgrenadierkompanie Always Attacks against an Infantry Company (see page 257 of the rulebook).

If you are fielding a Volksgrenadier Scout Platoon, and it is deployed on table at the start of the game, the Volksgrenadier Scout Platoon and all of the Volksgrenadier Sturm, Volksgrenadier Schützen, and Volks Pioneer Platoons in your force may make a Spearhead Deployment move (see page 261 of the rulebook).

VOLKSGRENADIERKOMPANIE

PEOPLE'S GRENADIER COMPANY

INFANTRY COMPANY

MOTIVATION AND SKILL

Platoons from 277. Volksgrenadierdivision *(marked)* *are rated* **Reluctant Trained.**

Platoons from 12. Volksgrenadierdivision *(marked)* *are rated* **Confident Veteran.**

277. VOLKSGRENADIER-DIVISION

RELUCTANT	CONSCRIPT
CONFIDENT	TRAINED
FEARLESS	VETERAN

12. VOLKSGRENADIER-DIVISION

RELUCTANT	CONSCRIPT
CONFIDENT	TRAINED
FEARLESS	VETERAN

You must field one platoon from each box shaded black and may field one platoon from each box shaded grey.

Your force must be either 277. Volksgrenadierdivision (marked) or 12. Volksgrenadierdivision (marked). If a platoon has either of these symbols, you can only take equipment and options marked with your selected symbol. If the platoon does not have either symbol, you can take equipment and options marked with any symbol, or no symbol.

HEADQUARTERS

- HEADQUARTERS: Volksgrenadier-kompanie HQ — 47

COMBAT PLATOONS

- INFANTRY: Volksgrenadier Sturm Platoon — 47
- INFANTRY: Volksgrenadier Schützen Platoon — 47
- INFANTRY: Volksgrenadier Sturm Platoon — 47

WEAPONS PLATOONS

- ARTILLERY: Volks Mortar Platoon — 48
- ARTILLERY: Volks Infantry Gun Platoon — 48
- MACHINE-GUNS: Volks Machine-gun Platoon — 49

REGIMENTAL SUPPORT PLATOONS

- ARTILLERY: Volks Heavy Mortar Platoon — 49
- ARTILLERY: Volks Infantry Gun Platoon — 48
- RECONNAISSANCE: Volksgrenadier Scout Platoon — 49; Skorzany Commando Group — 72

DIVISIONAL SUPPORT PLATOONS

- ARMOUR: Volks Tank-hunter Platoon — 50
- ARMOUR: Panzer Platoon — 69; Schwere Panzerjäger Platoon — 67; Sturmpanzer Platoon — 69; Tank-hunter Platoon — 69; Heavy Assault Howitzer Platoon — 71
- ANTI-TANK: Volks Anti-tank Gun Platoon — 50
- ANTI-AIRCRAFT: Volks Anti-aircraft Gun Platoon — 51
- INFANTRY: Panzergrenadier Platoon — 63; Volksgrenadier Sturm Platoon — 47; Volksgrenadier Schützen Platoon — 47
- INFANTRY: Volks Pioneer Platoon — 51
- ARTILLERY: Volks Light Artillery Battery — 52; Volks Heavy Artillery Battery — 52; Volks Corps Artillery Battery — 52
- ARTILLERY: Volks Light Artillery Battery — 52; Volks Rocket Launcher Battery — 53
- AIRCRAFT: Air Support — 68

HEADQUARTERS

VOLKSGRENADIERKOMPANIE HQ

HEADQUARTERS

Company HQ	50 points	80 points
Add:		
3 Panzerschreck teams	+60 points	+75 points
2 Panzerschreck teams	+40 points	+50 points
1 Panzerschreck team	+20 points	+25 points

OPTION

- Add up to three Sniper teams for +50 points per team.

COMBAT PLATOONS

VOLKSGRENADIER STURM PLATOON

PLATOON

HQ Section with:		
2 Assault Squads and 1 MG Squad	160 points	270 points
2 Assault Squads	120 points	200 points
Add:		
1 Panzerschreck team	+20 points	+25 points

A *Volksgrenadierkompanie* (People's Grenadier Company) is made up of three platoons, two of which are *Sturm* (Assault) Platoons. Two squads are armed with StG44 (*Sturmgewehr 44*) assault rifles giving them great firepower on the move, while the last section is armed with two MG42 machine-guns for supporting fire They are also equipped with *Panzerfaust* and *Panzerschreck* anti-tank weapons.

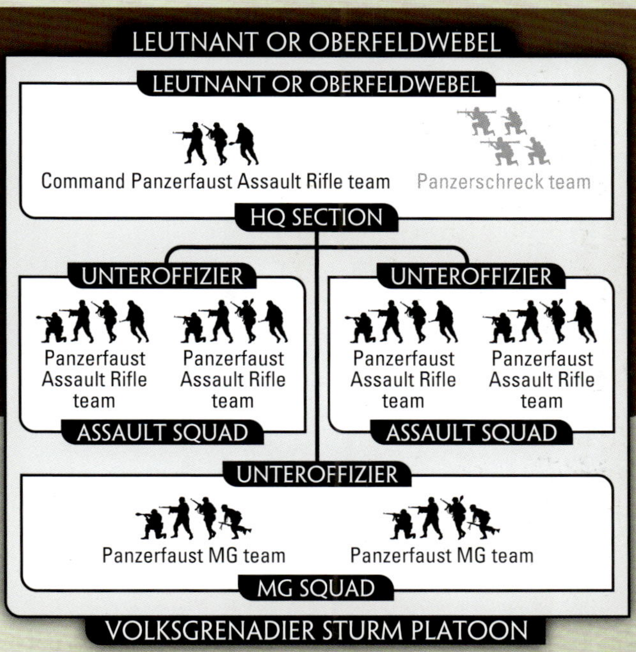

VOLKSGRENADIER SCHÜTZEN PLATOON

PLATOON

HQ Section with:		
3 Grenadier Squads	130 points	220 points
2 Grenadier Squads	90 points	155 points
Add:		
1 Panzerschreck team	+20 points	+25 points

OPTION

- Replace Command Panzerfaust Rifle/MG team with a Command Panzerfaust SMG team at no cost.

The third platoon of the company, the *Schützen* (Rifle) Platoon, is armed with K98 rifles and MG42 machine-guns. They provide support and longer ranged firepower to back-up the *Sturm* Platoons.

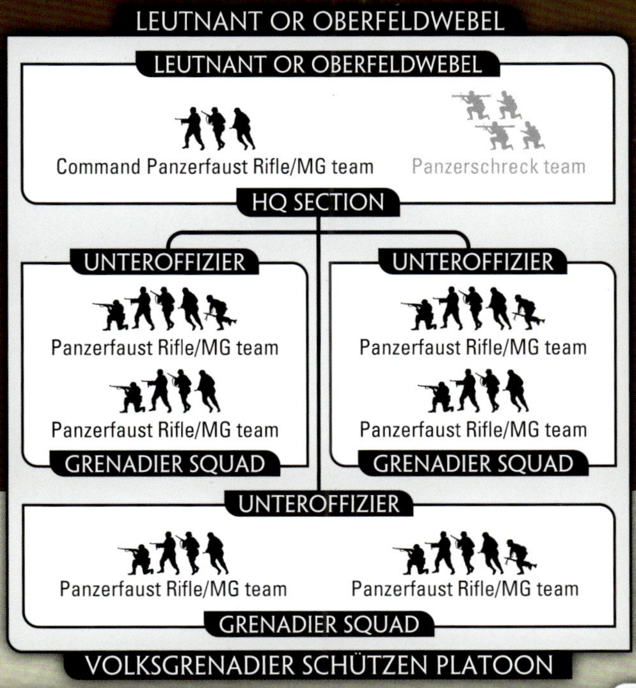

WEAPONS PLATOONS

VOLKS MORTAR PLATOON

PLATOON

HQ Section with:

3 Mortar Sections	130 points	180 points
2 Mortar Sections	90 points	125 points

OPTION

- Replace Command SMG team with Command Panzerfaust SMG team for +10 points.

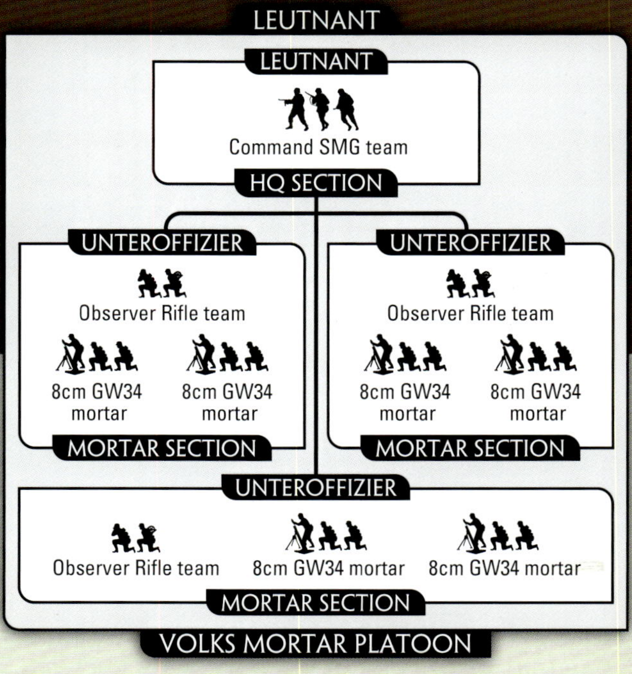

Mortars play an important role on both defence and attack. They can be quickly called on to deliver deadly fire on an attacking enemy, especially if they can catch the enemy out in the open as they approach the defensive lines.

On attack, they can deliver a barrage on a position to keep enemy heads down and reduce incoming fire on the attacking platoons. In direct support, they are ideal for pin-pointing troublesome guns or machine-guns and knocking them out.

VOLKS INFANTRY GUN PLATOON

PLATOON

HQ Section with:

4 7.5cm IG37	80 points	120 points
2 7.5cm IG37	45 points	65 points
4 7.5cm PaK50	95 points	140 points
2 7.5cm PaK50	50 points	75 points

OPTIONS

- Replace Command SMG team with Command Panzerfaust SMG team for +10 points.
- Add Horse-drawn limbers for +5 points for the platoon.

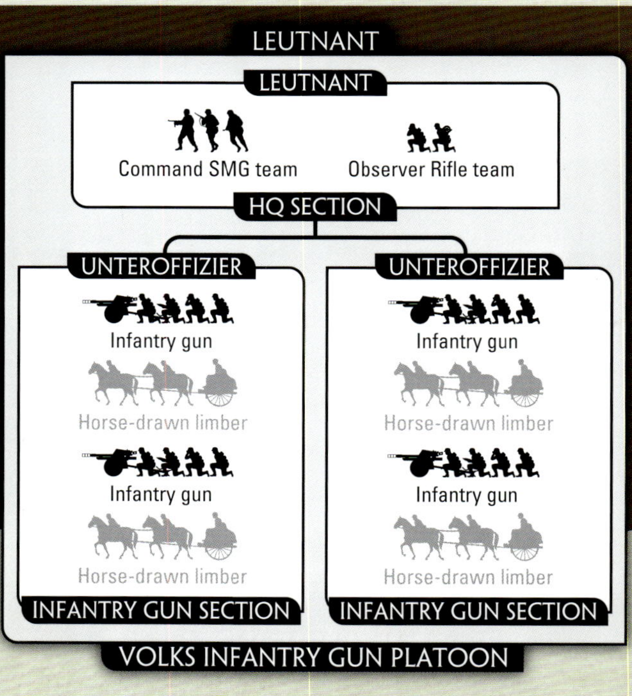

The *Volksgrenadier* divisions are among the first to receive a new generation of infantry guns. The 7.5cm leIG37 gun is a modernised gun on a light 3.7cm PaK36 trail. The 7.5cm PaK50 is a more sophisticated gun based on a cut down 7.5cm PaK40 on a light 5cm PaK38 trail. It has almost the same power as the 7.5cm PaK40, but uses a large muzzle brake to prevent the weapon being ripped apart every time it fires.

Volks Machine-gun Platoon

Platoon

HQ Section with:

2 Machine-gun Sections	90 points	135 points

Option

- Replace Command SMG team with Command Panzerfaust SMG team for +10 points.

Volks Machine-gun Platoons may make Combat Attachments to Volksgrenadier Sturm and Volksgrenadier Schützen Platoons.

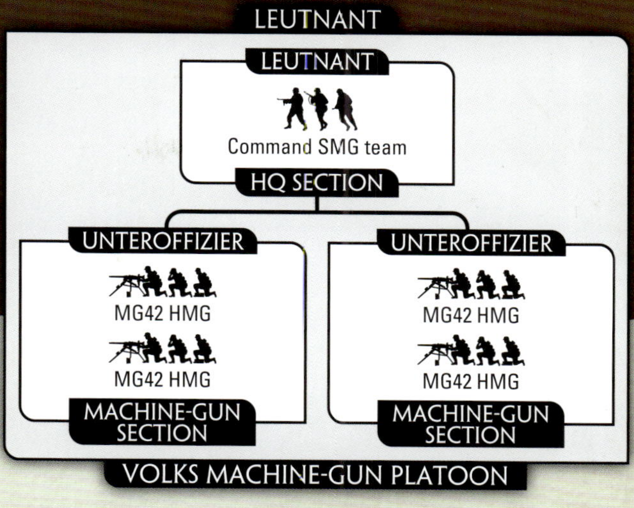

REGIMENTAL SUPPORT PLATOONS

Volks Heavy Mortar Platoon

Platoon

HQ Section with:

4 12cm sGW43	105 points	155 points

Option

- Add Kübelwagen jeeps and 3-ton trucks for +5 points for the platoon.

The 12cm sGW43 heavy mortar is based on the Soviet design used on the Eastern Front. It is light weight, which provides a high degree of mobility.

Though the mortar is light, it is still capable of delivering impressively destructive round. It gives the *Volksgrenadierkompanie* readily available heavy firepower to either halt an enemy assault or soften up enemy positions.

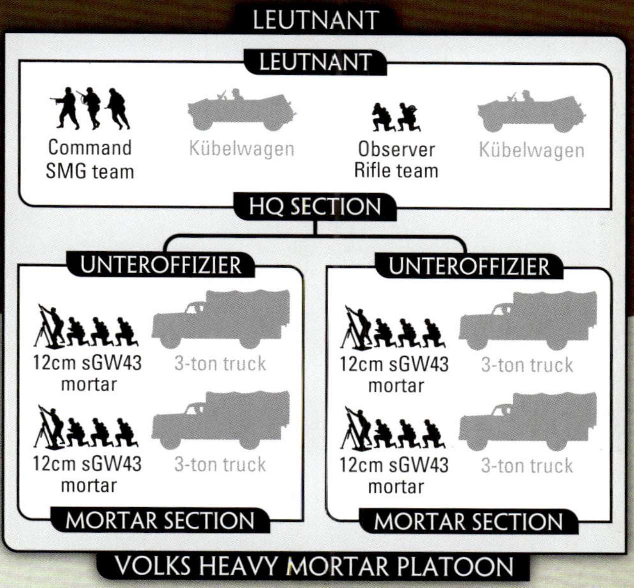

Volksgrenadier Scout Platoon

Platoon

HQ Section with:

2 Scout Squads	70 points	115 points
1 Scout Squad	40 points	70 points

Options

- Replace Command Rifle team with a Command Panzerfaust SMG team for +15 points or a Command Panzerfaust Assault Rifle team for +25 points.
- Replace all Rifle teams with Assault Rifle teams for +15 points per team.

Volksgrenader Scout Platoons are Reconnaissance Platoons and use the Stosstaktik special rule on page 45.

During the attacks in the Ardennes, the scouts played an important role in discovering the weakest points in the American defences. Using the cover of darkness they could often infiltrate themselves and accompanying *Volksgrenadier* platoons into the enemy lines to take them by surprise.

VOLKSGRENADIER DIVISIONAL SUPPORT PLATOONS

Volks Tank-hunter Platoon

Platoon

4 Hetzer	235 points	340 points
3 Hetzer	175 points	255 points
2 Hetzer	115 points	170 points
4 StuG G or StuG IV	260 points	380 points
3 StuG G or StuG IV	195 points	285 points
2 StuG G or StuG IV	130 points	190 points
4 StuG G (late)	265 points	395 points
3 StuG G (late)	200 points	295 points
2 StuG G (late)	135 points	195 points

Each *Volksgrenadierdivision* has a company of tank-hunters. They were usually either *Panzerjäger 38(t)* Hetzers, like the *277. Volksgrenadierdivision* and *26. Volksgrenadierdivision*, or StuG assault guns, like *12. Volksgrenadierdivision*.

Volks Anti-tank Gun Platoon

Platoon

HQ Section with:

3 7.5cm PaK40	105 points	155 points
3 7.62cm PaK36(r)	90 points	125 points

Options

- Add Kübelwagen jeep and 3-ton trucks for +5 points for the platoon.
- Replace 3-ton trucks with RSO tractors at no cost.

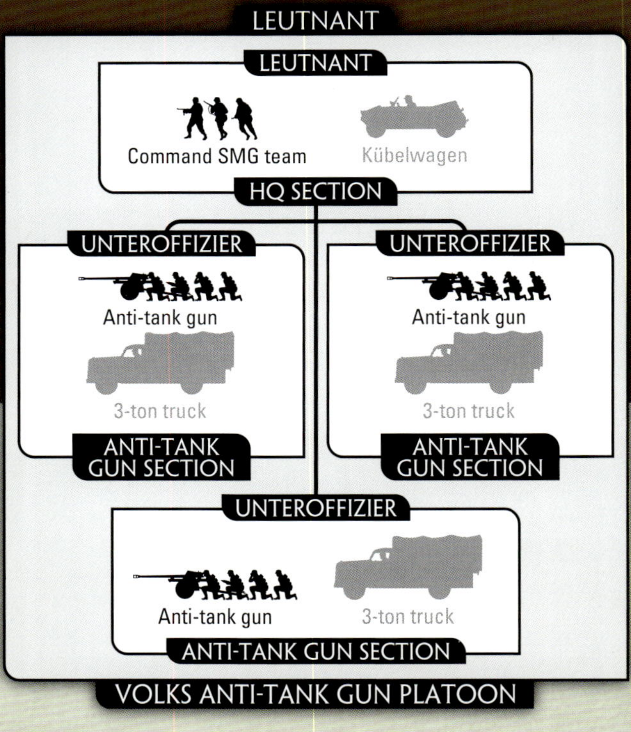

As well as tank-hunters or assault guns, the *Panzerjäger Abteilung* (anti-tank battalion) of each *Volksgrenadierdivision* also has towed anti-tank guns. Most are armed with 7.5cm PaK40 guns, but a few divisions have the older 7.62cm PaK36(r) gun.

The 7.5cm PaK40 gun has become the standard anti-tank gun of the Germans as lighter weapons have been retired. It is hard hitting and can deal with just about any tank the western allies can put against it.

Volks Pioneer Platoon

Platoon

HQ Section with:

3 Pioneer Squads	160 points	275 points
2 Pioneer Squads	115 points	190 points

Options

- Replace Command Panzerfaust Pioneer Rifle team with Command Panzerfaust Pioneer SMG team for +5 points.
- Add Pioneer Supply horse-drawn wagon for +20 points or Pioneer Supply 3-ton truck or RSO tractor for +25 points.

You may replace up to one Panzerfaust Pioneer Rifle team per Pioneer Squad with a Flame-thrower team at the start of the game before deployment.

A *Volksgrenadierdivision* has its own pioneers for laying and clearing mines and other obstacles. They are also armed with explosive charges and flame-throwers, which are ideal in assaults against fortified positions and bunkers.

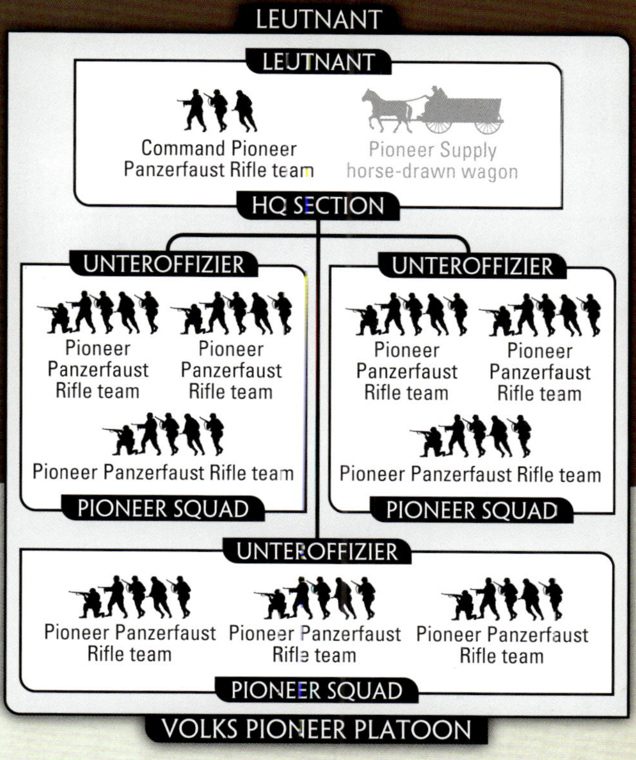

Volks Anti-aircraft Gun Platoon

Platoon

HQ Section with:

3 3.7cm FlaK43	65 points	100 points

Option

- Add 3-ton trucks for +5 points for the platoon.

With the dominance of the skies by the western Allies' fighter-bomber aircraft, the *Volksgrenadier* divisions have been given 3.7cm FlaK43 anti-aircraft guns. With better hitting power and longer range than the old 2cm FlaK38 gun, these new guns provide the infantry with better protection from air attacks.

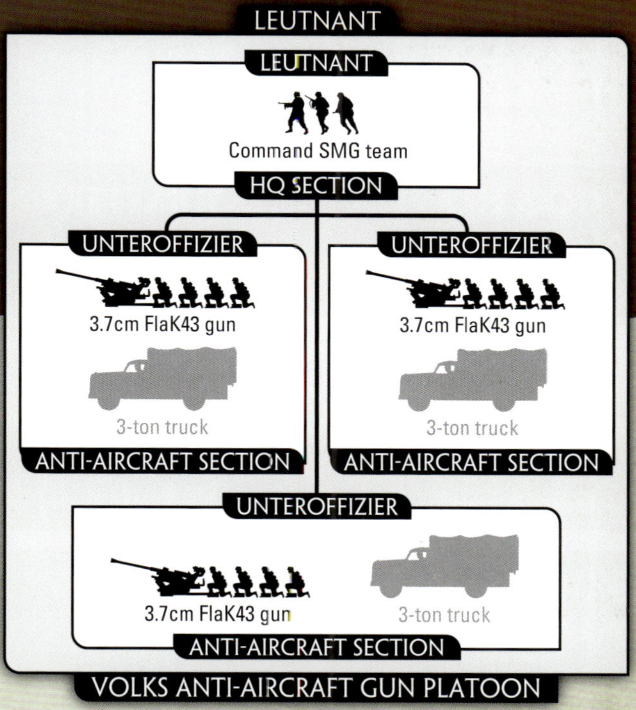

Volks Light Artillery Battery

Platoon

HQ Section with:

6 10.5cm leFH18/40	195 points	290 points
3 10.5cm leFH18/40	105 points	155 points
6 7.5cm FK40	225 points	335 points
3 7.5cm FK40	120 points	180 points

Option

- Add horse-drawn wagon and limbers for +5 points for the battery.

You must purchase all the guns from one Gun Section before adding any teams from another Gun Section.

Light Artillery Batteries may not be placed from Ambush within 16"/40cm of enemy teams.

The artillery of a *Volksgrenadierdivision* was made up of light and heavy artillery. Unusually, the light artillery could be the standard 10.5cm leFH18/40 field howitzer or the 7.5cm FK40 field gun. The FK40 was simply a 7.5cm PaK anti-tank gun used for artillery work with trained artillery crews, but it could still be just as deadly against tanks.

Volks Heavy Artillery Battery

Platoon

HQ Section with:

6 15cm sFH18	290 points	425 points
3 15cm sFH18	155 points	225 points

Option

- Add horse-drawn wagon and limbers for +5 points for the battery.

Volks Corps Artillery Battery

Platoon

HQ Section with:

6 15.2cm KH433(r)	315 points
3 15.2cm KH433(r)	165 points
6 8.8cm FK43 (PaK43)	425 points
3 8.8cm FK43 (PaK43)	225 points

Option

- Add horse-drawn wagon and limbers for +5 points for the battery.

A Volks Corps Artillery Battery is rated **Reluctant Trained.**

RELUCTANT | TRAINED

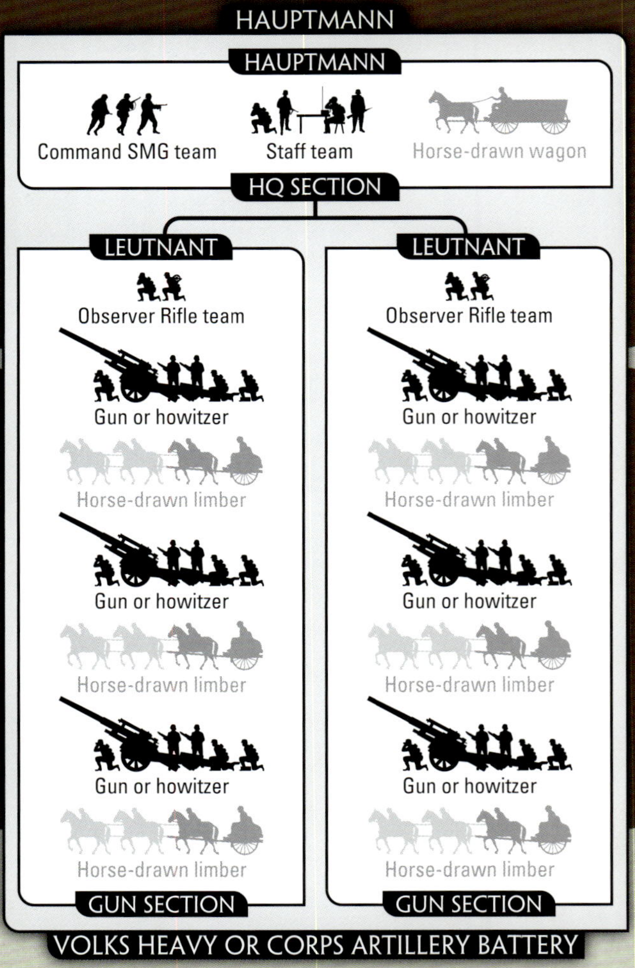

Volks Rocket Launcher Battery

Platoon

HQ Section with:

6 15cm NW41	135 points
3 15cm NW41	70 points
6 21cm NW42	160 points
3 21cm NW42	85 points
6 30cm NW42	195 points
3 30cm NW42	100 points

Option

- Add Kübelwagen jeep and Sd Kfz 11 half-tracks +5 points for the battery.

A Volks Rocket Launcher Battery is rated **Reluctant Trained.**

RELUCTANT | TRAINED

Super-heavy Rockets

The massive 30cm rockets used by the NW42 rocket launchers are slow to load and therefore the full battery is not always available to fire on command. A Volks Rocket Launcher Battery equipped with 30cm NW42 rocket launchers uses the Super-heavy Rockets special rule.

At the start of the game, place a Full Salvo marker with a platoon equipped with 30cm NW42 Super-heavy Rockets. Remove this marker after firing an Artillery Bombardment.

If a platoon with Super-heavy Rockets does not have a Full Salvo marker when it fires an Artillery Bombardment, roll a Skill Test for each Rocket Launcher able to fire in the Bombardment. Only those that pass the Skill Test can fire as part of the Bombardment.

Place a Full Salvo marker on a platoon with Super-heavy Rockets at the end of any Shooting Step in which every Rocket Launcher in the platoon was able to fire an Artillery Bombardment, but none did so.

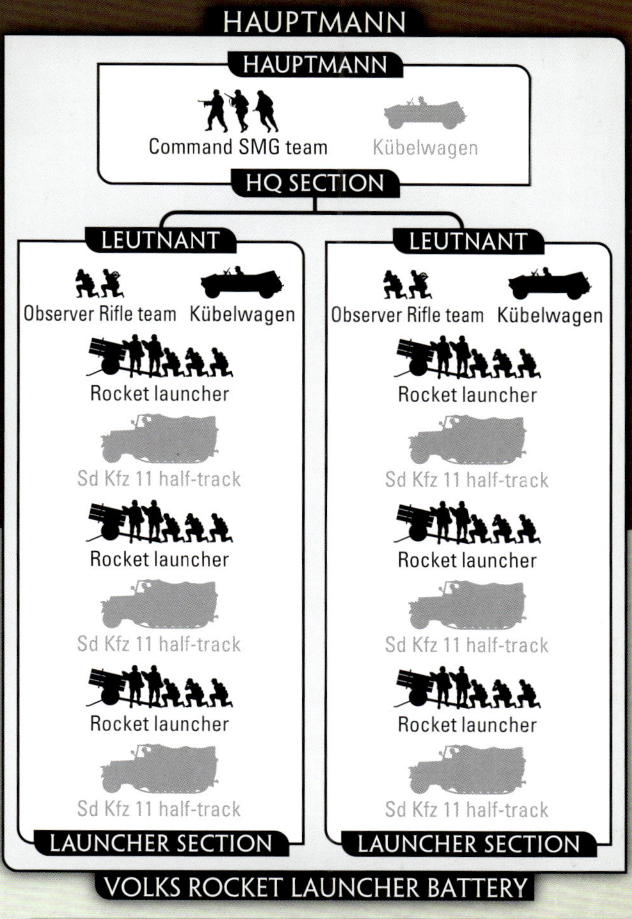

You must purchase all the rocket launchers from one Launcher Section before adding any teams from another Launcher Section.

Five *Volks-Werfer Brigaden* (People's Rocket Launcher Brigades) supported the German forces during the Ardennes offensive. Each brigade consisted of two regiments with six batteries of 15cm NW41 rocket launchers and three heavy batteries of either 21cm NW42 or 30cm NW42 rocket launchers, depending on if it was the first or second battalion.

The biggest advantage of *Nebelwerfer* rocket launchers is their ability to saturate a wide area quickly with high-explosive rounds. They are ideal for pinning down an enemy position while attacking forces approach for the assault.

OPERATION NORDWIND
HITLER'S LAST OFFENSIVE IN THE WEST

Unternehmen Nordwind (Operation North Wind) was the last major offensive launched by the Germans in the west. With the Ardennes offensive (*Unternehmen Wacht am Rhein*) raging to the north, the Americans had redirected reserves to halt the German armoured thrust towards the Meuse River and Antwerp. Hitler decided it was time to unleash another offensive on the American and French forces in Alsace-Lorraine. In a meeting with his divisional commanders on 28 December 1944, Hitler made the objectives of *Nordwind* clear:

> *'...It is a matter of destroying and exterminating the enemy forces wherever we find them. The question of liberating all of Alsace at this time is not involved either. That would be very nice, the impression on the German people would be immeasurable, the impression on the world decisive... But that is not important. It is more important, as I said before, to destroy his manpower.'*

Hitler wanted to destroy the US Seventh and French First Armies. By breaking through these two armies, the Germans would then turn their attention to the rear of the US Third Army. Like many of Hitler's offensives in the later stages of the war, the objectives were ambitious to say the least.

THE OFFENSIVE BEGINS

On 31 December two *Heersgruppen* (Army Groups) were committed to *Nordwind*. *Heersgruppe G* and *Heersgruppe Oberrhein* (Army Group Upper Rhine) struck the thinly stretched US Seventh Army. Initially, the assault was led by *XIII SS-Armeekorps* of *Heersgruppe G*. Spearheaded by *17. SS-Panzergrenadierdivision,* the corps broke through the US 44th Infantry Division east of Sarreguemines. However, US intelligence had predicted the attack would come in this sector and reinforcements pushed *17. SS* back. The attacks further east by *559.* and *361. Volksgrenadier* divisions were more successful and they had soon closed on Wingen-sur-Moder. There they met the US 45th Infantry Division who fought back hard, halting further advance.

MAGINOT LINE

On 3 January, with *XXII SS-Armeekorps* pushed back to its starting positions of 31 December, the expected exploitation in this sector was called off. The reserve *21. Panzerdivision* and *25. Panzergrenadierdivision* were moved to *LXXXX Korps*. The two divisions began advancing from Wissembourg

OPERATION NORDWIND, INITIAL ATTACKS

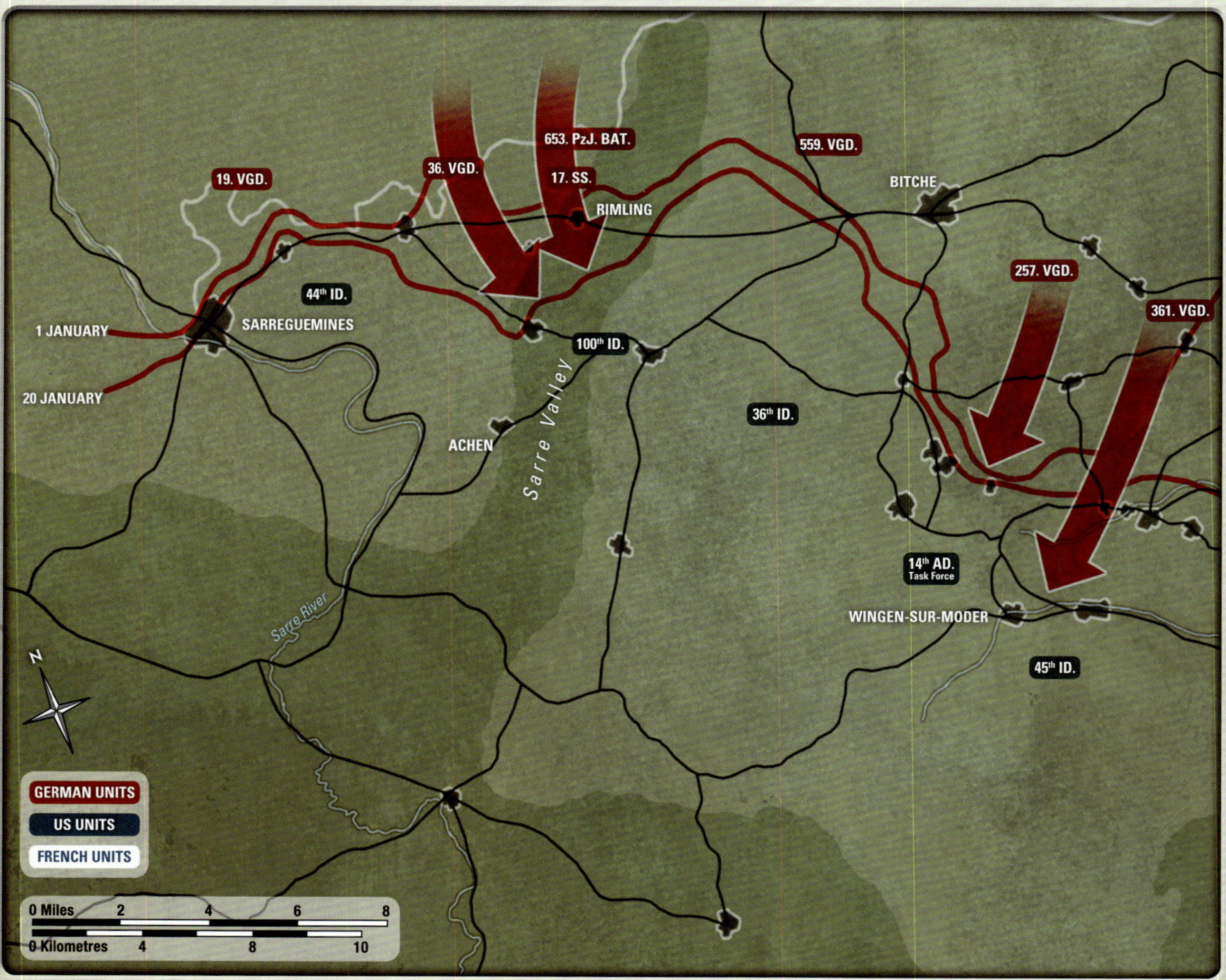

towards Haguenau while American forces fell back towards the Moder River through the Maginot Line. The Maginot fortified line had been built by the French in the 1930s. Many of the guns had been stripped out by the Germans to defend the Atlantic Wall, but the Americans had made use of mines, wire and the bunkers to anchor their defensive positions.

21. Panzerdivision and *25. Panzergrenadierdivision* assaulted the Maginot Line on 6 January 1945. They broke through and moved on to Rittershoffen and Hatten. They engaged in a savage battle for the two villages until 19 January, when the Americans were finally pushed out.

Strasbourg

Meanwhile, on 3 January *Heersgruppe Oberrhein*, under the command of *Reichsführer* Heinrich Himmler, entered the battle. *553. Volksgrenadierdivision* assaulted across the Rhine at Gambsheim towards Strasbourg. Further south, the Nineteenth Army in the Colmar Pocket launched an attack towards Strasbourg on 7 January with the veteran *198. Infanteriedivision, 106. Panzer Brigade* and Jagdpanther tank-hunters from *654. Schwere Panzerjägerabteilung*. Both thrusts were halted by the French First Army, many of whom were veterans of the campaigns in Italy.

The Allies Strike Back

The Allies launched a counterattack on 1 February across the front. Despite fighting gamely against the Franco-American onslaught, the seven weakened divisions of the Colmar Pocket were overwhelmed. The pocket fell and the remnants of the German Colmar forces were pushed across the Rhine by 9 February. The Germans were also pushed back at Gambsheim. The Germans held some of their gains in the Haguenau sector, even if the city itself was contested. The villages of Rittershoffen and Hatten were not retaken until the Allied offensive into the Saar in March 1945.

OPERATION NORDWIND, ATTACKS TOWARDS HAGUENAU

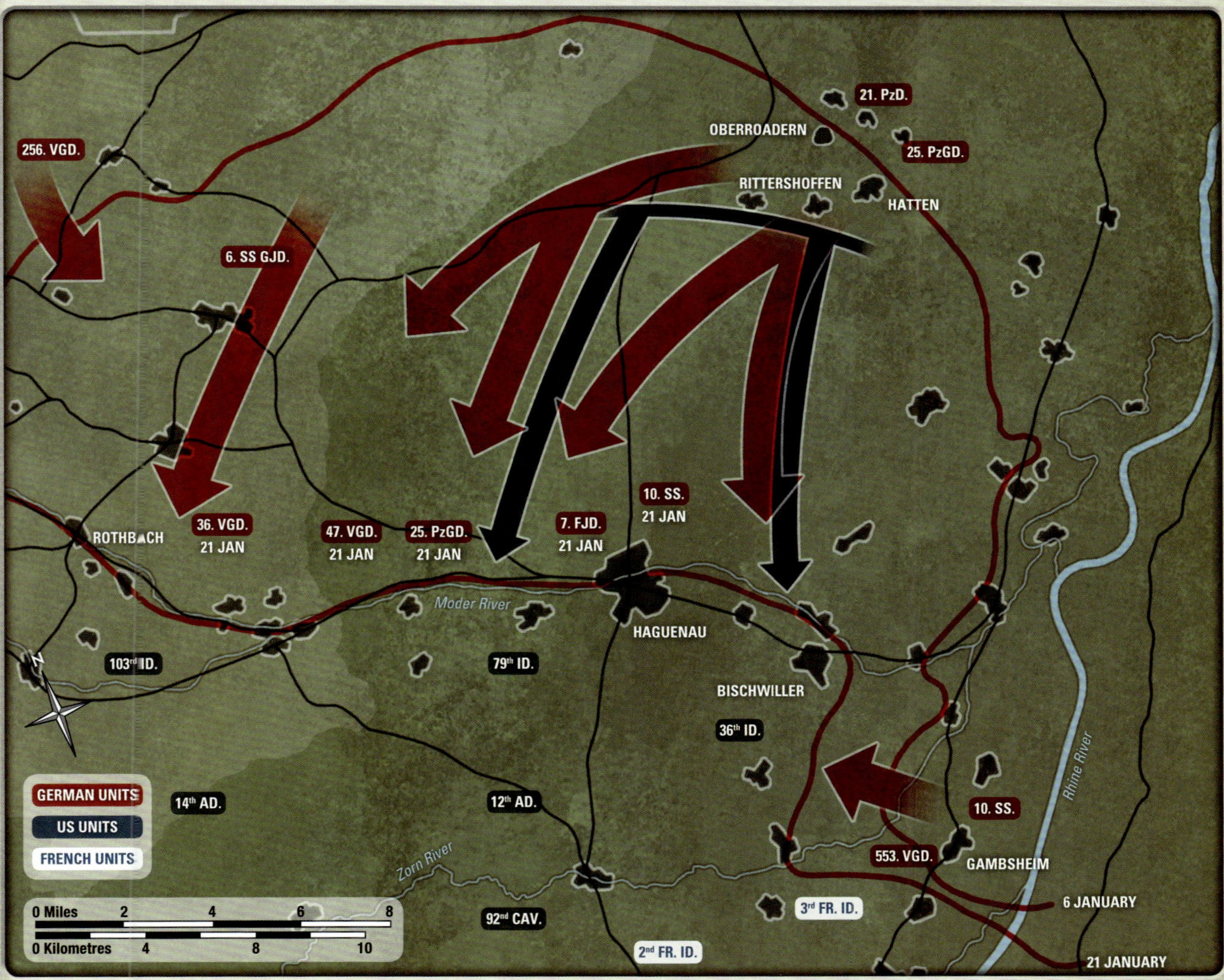

25. PANZERGRENADIERDIVISION

25. Panzergrenadierdivision was formed from the motorised *25. Infanteriedivision* in June 1943. It fought on the Russian Front until it was encircled at Minsk in July 1944. Reformed from the remnants of the division in October 1944, initially as *107. Panzerbrigade*, it was expanded to a full division and renamed the *25. Panzergrenadierdivision* once more.

The division was then moved to the Germany-Luxembourg-France border area near Sierck-les-Bains where it fought delaying actions against Patton's Third Army until December. It was then moved to the town of Bitche in the Vosges region of Alsace-Lorraine where it fought to hold the US Seventh Army.

Preparations for Nordwind

As part of the planning for the Ardennes offensive, it was anticipated that a local offensive into Alsace-Lorraine would be called for to relieve pressure in the Ardennes. With the US Seventh Army's pressure on *Heersgruppe G* dying down by 20 December, it gave the Germans an opportunity to start preparing for such an attack by pulling *21. Panzerdivision* and *25. Panzergrenadierdivision* out of the line to rest and refit in preparation. The two divisions were joined by *6. SS-Gebirgsdivision 'Nord'* as the army group's mobile reserve.

The *25. Panzergrenadierdivision* spent the initial phases of the offensive in reserve. Its two armoured units were equipped with Panther tanks and StuG assault guns, but only had about a third of their authorised strength before some new vehicles were delivered. For the operation the division was allocated *352. Panzerflammkompanie* armed with new *Flammpanzer 38(t) Hetzer* flame-thrower tanks.

Like the Ardennes offensive, the infantry were called on to break through the US lines to allow the panzer troops to punch through the gaps. However, *25. Panzergrenadierdivision* was to wait until *XIII SS-Armeekorps* (13th SS Corps) had broken through the Saar Valley so they could thrust into the open country to the west. The failure of *XIII SS-Armeekorps* necessitated moving the focus of the attack to another sector. The reserve panzer troops were moved east to Rothbach so they could break into the open country east of Haguenau.

Maginot Line

After waiting to attack since 31 December, *25. Panzergrenadierdivision* was finally unleashed on 6 January 1945. They fought alongside the severely under strength *21. Panzerdivision* as the combined *Kampfgruppe Feuchtinger*. They moved out from Bienwald at 0200 hours with the aim of taking several Maginot Line forts by surprise. The opposition from Task Force Linden (42nd Infantry Division) was light, but the advance was slowed by minefields. The following day, the push continued into the Maginot Line defences at Oberroedern held by the US III Battalion, 313th Infantry Regiment, 79th Infantry Division. Engineers and tanks finally cleared a gap through the Maginot Line just north of Oberroedern on 8 January and *25. Panzergrenadierdivision* began pushing through the defences towards the towns of Hatten and Rittershoffen that evening. Fighting through wire and minefields, the panzergrenadiers were able to knock out several bunkers with the aid of StuG assault guns and *Flammpanzer 38(t) Hetzer* flame-thrower tanks while under intense artillery fire.

Hatten and Rittershoffen

The division's *119. Panzergrenadierregiment* began to assault Hatten at dawn on 9 January, clashing with the infantry of Task Force Linden (242nd Infantry Regiment). Meanwhile *Kampfgruppe Huss* (*25. Panzergrenadierdivision*) sent a battalion of half-track mounted panzergrenadiers with tank support to attack the town from the other flank. The Americans moved up Combat Command A from the 14th Armored Division and in the blinding snow the Germans did not spot the American tanks. *Kampfgruppe Huss* soon retreated under heavy tank and artillery fire. The Americans then counterattacked and relieved the Hatten defenders.

The positions around Hatten and Rittershoffen exchanged hands several times in the course of a week of hard fought battles. The *119. Panzergrenadierregiment* held the Americans while the rest of *21. Panzerdivision* and *25. Panzergrenadiedivision* fought to widen the gap in the Maginot Line. The two American battalions holding Hatten and Rittershoffen clung to the rubble, holding out against the panzergrenadiers' concerted attacks. More armoured infantry and tanks from the US 14th Armored Division arrived and began an attempt to relieve Hatten, but the Germans held them off and made their own counterattacks.

On the night of 12/13 January troops of *7. Fallschirmjägerdivision* began arriving and finally tipped the balance. The Germans pushed the Americans out of the towns and, in the following few days, they held off further American counterattacks. Finally, on the evening of 19 January *47. Volksgrenadierdivision* arrived to take over the defence of Hatten and Rittershoffen, allowing *21. Panzerdivision* and *25. Panzergrenadierdivision* to retire on 20 January. The Americans withdrew to the Moder River the following day.

In late January 1945 *25. Panzergrenadierdivision* was transferred back to the Eastern Front by train. For the remainder of the war they fought against the Soviet Drive towards Berlin, finally surrendering to the Americans at Radelübbe in May 1945.

17. SS-PANZERGRENADIERDIVISION

17. SS-Panzergrenadierdivision was formed near Poitiers, France from October 1943. Many of the recruits were *Volksdeutch* (ethnic Germans) from Romania. The division was granted the title *'Götz von Berlichingen'* after the 15th century German knight and mercenary Gottfried von Berlichingen. The divisional symbol was an iron fist in a shield, taken from Berlichingen's iron prosthetic forearm, which he wore after losing his original during a siege.

NORMANDY AND FRANCE

After the D-Day landings on 6 June, *17. SS-Panzergrenadierdivision* was ordered from their garrison in Thouars, France to Normandy with orders to halt the invasion. For the next month, *17. SS-Panzergrenadierdivision* was involved in fighting in the bocage country around St. Lô and Coutances, where the division suffered heavy losses. The division escaped entrapment in the Falaise Pocket and remained in constant contact with pursuing US forces until the end of August.

The remains of the division were finally withdrawn to Metz in September to rest and refit. Their rest was cut short and they were thrown back into battle to destroy the US bridgehead over the River Moselle. After heavy fighting, the division was forced to fall back to the Saar Region. On 8 November a US bombing raid knocked out their divisional command post and their senior officers. With the unit in tatters and no command, they were withdrawn for refitting.

REBUILD AND OPERATION NORDWIND

When the division was withdrawn to the Maginot Line in mid November, as they only had 4000 men and 20 armoured vehicles after two months of hard fighting. During the refit, the *panzergrenadier* regiments were brought up to full strength with *Volksdeutch* replacements, though the quality of the replacements was varied. The division's original surviving StuG IV assault guns were supplemented with new StuG III G assault guns. They joined the *XIII SS-Korps* to participate in Operation *Nordwind*, receiving additional support from *Panzerflammkompanie 353* and a company of Jagdtiger tank-hunters from *653. Schwere Panzerjagerabteilung*.

The attack began on 31 December 1944 at 2330 hours from positions in front of the village of Rimling. *17. SS-Panzergrenadierdivision* attacked as part of *Sturmgruppe 1* (assault group 1) of *XIII SS-Korps*, flanked on their right by *36. Volksgrenadierdivision* and left by *559. Volksgrenadierdivision*. They faced the American 44th and 100th Infantry Divisions, defending deep positions in the Sarre Valley. The ground was covered in about a foot (30cm) of snow, making road conditions difficult for their armoured support. The Jagdtiger tank-hunters saw their first action during the battle at Rimling in support of *17. SS-Panzergrenadierdivision* (see page 58).

Fighting was bloody and casualties on both sides were high. *17. SS-Panzergrenadierdivision* made little headway against stubborn US defence in freezing weather conditions. The division also suffered from inexperience among its higher commanders. Eventually, the whole divisional staff was relieved of command on 3 January and replaced with more experienced *Heer* (Army) officers. On 10 January the remaining senior SS officer, the division commander *SS-Standartenführer* Hans Lingner, was captured by a US patrol after his car overturned on a slippery road. Command of the division was taken over by *Heer* officer *Oberst* Gerhard Lindner.

With little success in their sector, the focus of Operation *Nordwind* moved east to the region around Haguenau. Though the offensive halted in their sector, *17. SS-Panzergrenadierdivision* continued to hold off US counterattacks and probe the enemy lines until the end of the operation on 25 January.

The division continued to fight in the defence of the western frontier of Germany until it was finally forced to abandon its vehicles and cross the Rhine in March 1945. The division was captured at Moosburg, Germany in April 1945.

653. SCHWERE PANZERJÄGERABTEILUNG

653RD HEAVY TANK-HUNTER BATTALION

The first unit to receive the *Jagdtiger* heavy tank-hunter was *653. Schwere Panzerjägerabteilung*. The battalion had already made a name for itself on the Eastern Front operating the *Elefant* heavy tank-hunter. Crews of *653. Schwere Panzerjägerabteilung* began training with the *Jagdtiger* in September 1944. Later that month the 1st and 3rd Companies moved to Linz, Austria to continue training at the Nibelungen Works where the *Jagdtiger* was being produced. Driving and gunnery training continued until 17 November 1944.

On 7 December the 1st Company left the training area with seven tank-hunters with the intention of taking part in the upcoming Ardennes offensive. Unfortunately, bomb damage to the rail network meant the unit never made it in time to take part in the operation.

OPERATION NORDWIND

It wouldn't be until the launch of Operation *Nordwind* on 31 December that the *Jagdtiger* tank-hunters would see their first action. Two tank-hunters of 3rd Company arrived in time to take part in the beginning of the offensive, while the rest of the battalion was still in the area of the Mosel River north of Lorraine. They were subordinated to *17. SS-Panzergrenadierdivision*. Four more *Jagdtiger* tank-hunters from the 1st Company joined them on 4 January.

On 9 January three *Jagdtiger* tank-hunters took part in an action at Rimling. Supporting *17. SS-Panzergrenadierdivision* infantry, they attacked the American occupied town. The tank-hunters arrived at the town around midday and positioned themselves on a slope on the outskirts of the town. The tank-hunters were spread out, about 250 to 300 meters apart. There they waited for their support infantry to arrive for about two hours. Suddenly, the enemy opened fire on them. The direction and location of the enemy fire couldn't be determined, so the *Jagdtiger* tank-hunters backed off. However, the Americans had worked around their flank and a shot from either a Bazooka or a M36 Tank Destroyer (accounts differ) caused an ammunition explosion on *Jagdtiger* 134, knocking out the vehicle and killing the entire crew.

A few days later, another engagement at Rimling was fought near the mill at about midnight, once again some *Jagdtiger* tank-hunters supported *17. SS-Panzergrenadierdivision*. The *Jagdtiger* tank-hunters made such good targets against the white snow during the day, that the crews had begun using them at night. They had learnt that the biggest threat to their massive tank-hunters was artillery and aircraft. Key to using these vehicles at night was good infantry support and, at the second engagement at Rimling, it was the infantry that protected them from counterattacking American infantry. On 11 January the *Jagdtiger* companies were withdrawn for rest.

As the focus of *Nordwind* switched to the Haguenau area, other elements of the battalion were committed to the fighting there. They fought in the area around Haguenau and Bischweiler, being use against bunkers on the Maginot Line in support of *10. SS-Panzerdivision 'Frundsburg'*. An official report noted that the *Jagdtiger* guns proved effective against bunkers with high-explosive rounds. *Jagdtiger* tank-hunters engaged four heavy bunkers and knocked out a heavy turret with two rounds on 18 February. American Sherman tanks counterattacked in the same area and the tank-hunters were able to repulse the enemy attack with a number of Sherman tanks set ablaze with only high-explosive rounds.

AFTER NORDWIND

By mid-February *653. Schwere Panzerjägerabteilung* was concentrated in the Haguenau Forest (north of Haguenau) covering the approaches to the West Wall. Throughout February, most of the *Jagdtiger* tank-hunters were kept in concealed positions awaiting possible Allied attacks into Germany.

On 14 March the order was given to withdraw from the positions around Haguenau. *653. Schwere Panzerjägerabteilung* was heavily engaged during the withdrawal and, once contact was broken with the enemy, they still continued to be harassed by P-47 fighter-bombers.

They fought the Americans around Weissenburg, Germany on 22 March, engaging US tank columns at long ranges and wreaking havoc amongst them. At Neustadt, three *Jagdtiger* tank-hunters destroyed 25 Sherman tanks and M10 tank destroyers without any losses. They continued their fighting retreat into April, finally ending the war in Austria at the *Jagdtiger* factory at the Nibelungen Works, surrendering on 7 May 1945.

Jagdtiger (12.8cm PaK44 auf Pazerjäger Tiger Ausf B, Sd Kfz 186)

Hull MG

Top Armour: 2

100%

Front Armour: 16

Movement: Slow Tank, Overloaded, Unreliable

Side Armour: 8

Weapon	*Range*	*ROF*	*AT*	*FP*	*Notes*
12.8cm PaK44 gun	*48"/120cm*	*2*	*17*	*2+*	*Hull mounted, Breakthrough gun.*

Crew:	6
Battle weight:	75 tonnes
Length:	10.65m
Width:	3.75m
Height:	2.94m
Engine:	700 HP
Main gun:	12.8cm
Armour:	40-250mm
Range on road:	170km
Max speed:	38km/h

The *Jagdtiger* was the largest and heaviest armoured fighting vehicle used by the German panzer forces during the war. Work on the *Jagdtiger* began in 1943 and Henschel, who already made the *Tiger* and *Königstiger* heavy tanks, submitted plans based on the *Königstiger*.

Two prototypes were produced, one using the original Henschel *Königstiger* suspension, the other based on the Porsche suspension similar to that of the *Elefant* tank-hunter. The Henschel design was selected, but due to slow production ten Porsche vehicles were also produced. A total of 77 *Jagdtiger* tank-hunters were manufactured by the end of the war.

The *Jagdtiger* tank-hunter had thick armour and was nearly invulnerable from the front and even from the side only better enemy anti-tank tank guns could penetrate it. Its main asset was its powerful 12.8cm PaK44 anti-tank gun. This gun was derived from the K81 divisional gun and could penetrate any tank the Allies had, often blowing smaller vehicles into fragments. However, its armour and gun made it very heavy, which put a massive strain on its drive train and engine. This led to reliability issues and made it difficult to get good numbers of them into combat at once.

Availability: The Jagdtiger heavy tank-hunter first went into action with *653. Schwere Panzerjägerabteilung* during Operation *Nordwind* in January 1945 and stayed in service until the end of the war.

Panzergrenadier Special Rules

A Panzergrenadierkompanie uses all the German special rules on pages 241 to 245 of the rulebook. In addition they also use the Concerted Effort and Under the Cloak of Darkness special rules.

A Schwere Panzerjägerkompanie uses all the German special rules on pages 241 to 245 of the rulebook, and the Concerted Effort special rule.

Concerted Effort

During the Ardennes and *Nordwind* offensives the *Heer*, *Waffen-SS* and *Luftwaffe* ground troops all worked together in a concerted effort to breakthrough the US lines.

Heer, Luftwaffe, and SS fight as one army, ignoring the Reich Divided special rule (see page 242 of the rulebook).

Under the Cloak of Darkness

To gain surprise *25. Panzergrenadierdivision* attacked the Maginot Line fortifications under the cover of darkness. Using their panzergrenadiers and pioneers, they closed with the fortifications using assault guns and *Flammpanzer 38(t)* flame-tanks in support.

A player commanding a Panzergrenadierkompanie (page 60) containing a Flammpanzer Platoon may elect to make a Night Attack in any Mission that they are facing a Fortified Company. If they do so, the game starts at night (see the Night Fighting rules on page 272 of the rulebook) and uses the Dawn rules on page 273 of the rulebook.

PANZERGRENADIERKOMPANIE

MOTORISED GRENADIER COMPANY

MECHANISED COMPANY

You must field one platoon from each box shaded black and may field one platoon from each box shaded grey.

Your force must be either Heer (marked) or SS (marked). If a platoon has either of these symbols, you can only take equipment and options marked with your selected symbol. If the platoon does not have either symbol, you can take equipment and options marked with any symbol, or no symbol.

HEADQUARTERS

HEADQUARTERS
- Panzergrenadierkompanie HQ — 61

COMBAT PLATOONS

INFANTRY
- Panzergrenadier Platoon — 61

INFANTRY
- Panzergrenadier Platoon — 61

INFANTRY
- Panzergrenadier Platoon — 61

INFANTRY
- Panzergrenadier Heavy Platoon — 62

WEAPONS PLATOONS

ARTILLERY
- Panzergrenadier Heavy Mortar Platoon — 63

ANTI-TANK
- Panzergrenadier Anti-tank Gun Platoon — 62

REGIMENTAL SUPPORT PLATOONS

ARTILLERY
- Panzergrenadier Infantry Gun Platoon — 64

INFANTRY
- Panzergrenadier Pioneer Platoon — 63

INFANTRY
- Armoured Panzergrenadier Platoon — 64

ANTI-AIRCRAFT
- Panzergrenadier FlaK Platoon — 63

DIVISIONAL SUPPORT PLATOONS

ARMOUR
- Panzergrenadier Panzer Platoon — 64
- Flammpanzer Platoon — 65

ARMOUR
- Panzergrenadier Panzer Platoon — 64
- Tank-hunter Platoon — 69
- Flammpanzer Platoon — 65

ARMOUR
- Schwere Panzerjäger Platoon — 67
- Heavy Assault Howitzer Platoon — 73

ANTI-TANK
- Panzergrenadier Anti-tank Gun Platoon — 62

RECONNAISSANCE
- Armoured Car Patrol — 65

ANTI-AIRCRAFT
- Panzergrenadier Anti-aircraft Gun Platoon — 65

INFANTRY
- Volksgrenadier Sturm Platoon — 47
- Volksgrenadier Schützen Platoon — 47

INFANTRY
- Panzergrenadier Pioneer Platoon — 63

ARTILLERY
- Mortorised Light Artillery Battery — 70
- Motorised Heavy Artillery Battery — 70
- Volks Corps Artillery Battery — 52

ARTILLERY
- Mortorised Light Artillery Battery — 70
- Volks Rocket Launcher Battery — 53

ANTI-AIRCRAFT
- Heavy Anti-aircraft Gun Platoon — 68

AIRCRAFT
- Air Support — 68

Motivation and Skill

A 25. Panzergrenadierdivision Panzergrenadierkompanie is rated as **Confident Veteran.**

A 17. SS-Panzergrenadierdivision SS-Panzergrenadierkompanie is rated as **Fearless Trained.**

25. Panzergrenadier-Division

RELUCTANT	CONSCRIPT
CONFIDENT	TRAINED
FEARLESS	**VETERAN**

17. SS-Panzer-Grenadierdivision

RELUCTANT	CONSCRIPT
CONFIDENT	**TRAINED**
FEARLESS	VETERAN

HEADQUARTERS

Panzergrenadierkompanie HQ

Headquarters

	25.	17. SS
Company HQ	65 points	60 points

Option

- Add Kübelwagen jeeps for +5 points for the HQ.

The main combat arm of a *Panzergrenadierdivision* is its motorised *Panzergrenadier* battalions and companies.

COMBAT PLATOONS

Panzergrenadier Platoon

Platoon

HQ Section with:

	25.	17. SS
3 Panzergrenadier Squads	245 points	225 points
2 Panzergrenadier Squads	175 points	160 points

Options

- Replace Command Panzerfaust MG team with Command Panzerfaust SMG team at no cost.
- Add Kübelwagen jeep and 3-ton trucks for +5 points for the platoon.

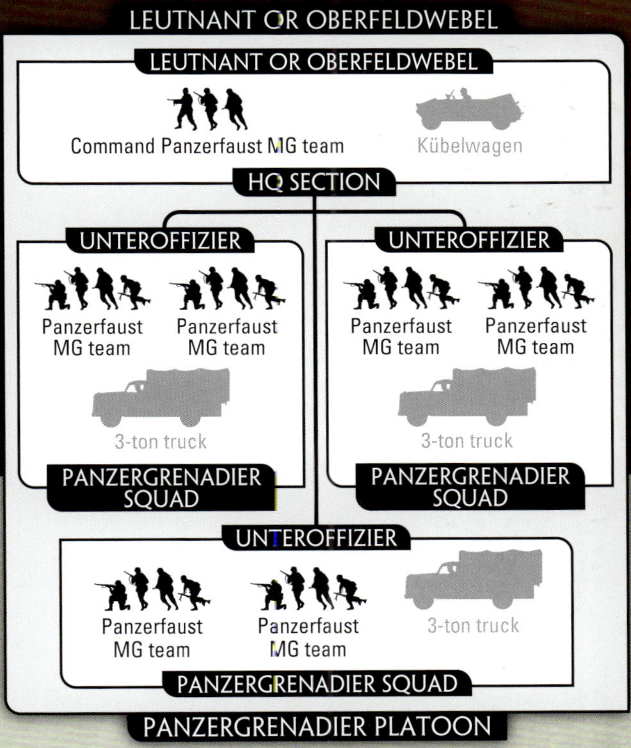

Each *Panzergrenadier* Platoon is equipped with MG42 machine-guns, K98 rifles and *Panzerfaust* anti-tank rockets. This provides them with amazing firepower, both on attack and defence.

Mounted in trucks, they are able to keep pace with their assault guns or panzers as they advance. They then dismount to fight, using their machine-guns to pin down the enemy before going in for the kill with rifles, grenades and entrenching tools.

Panzergrenadier Heavy Platoon

Platoon

HQ Section with:		
2 Machine-gun Sections	135 points	115 points
1 Machine-gun Section	70 points	60 points
No Machine-gun Sections	10 points	10 points
Add Mortar Section	+65 points	+55 points

Options

- Add Kübelwagen jeeps and 3-ton-trucks for +5 points for the platoon.
- Replace Command SMG team with Command Panzerfaust SMG team for +10 points.

A Panzergrenadier Heavy Platoon must have a Mortar Section if it has no Machine-gun Sections.

Panzergrenadier Heavy Platoons may make Combat Attachments to Panzergrenadier Platoons.

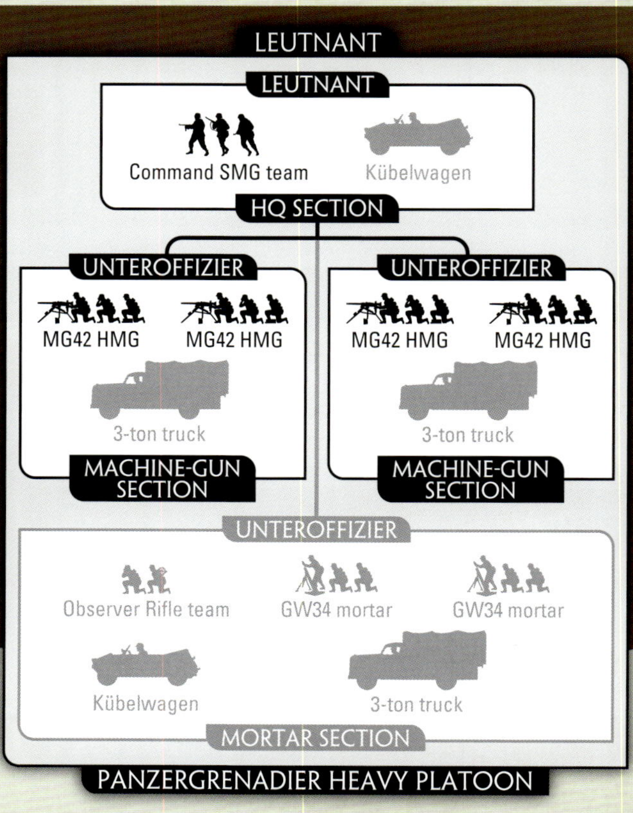

Weapons Platoons

Panzergrenadier Anti-tank Gun Platoon

Platoon

HQ Section with:		
3 7.5cm PaK40	155 points	135 points
2 7.5cm PaK40	105 points	90 points

Options

- Replace Command SMG team with Command Panzerfaust SMG team for +10 points.
- Add Kubelwagon jeep and 3-ton trucks for +5 points for the platoon.

The 7.5cm PaK40 anti-tank gun will deal with any armour the western allies care to throw against the panzergrenadiers. It has good range and excellent penetration, ideal for halting armoured counterattacks.

Each battalion had an anti-tank gun platoon to give direct support to the panzergrenadiers.

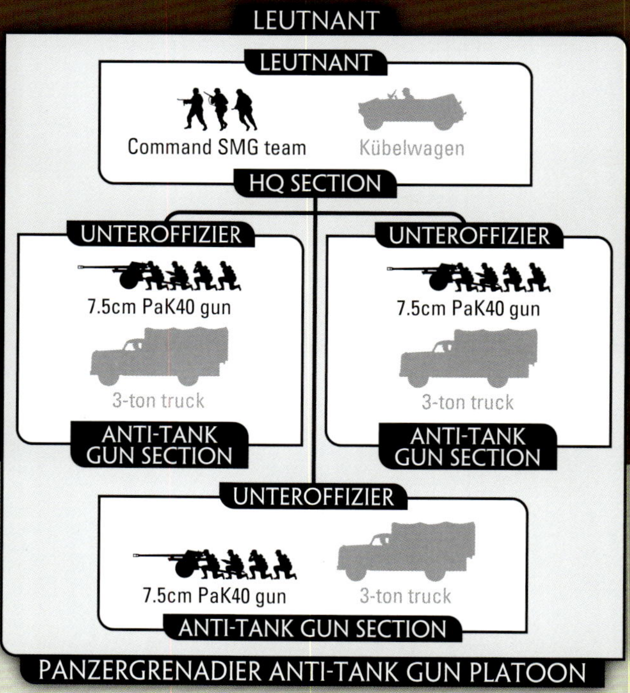

Panzergrenadier Heavy Mortar Platoon

Platoon

HQ Section with:

4 12cm sGW43	155 points	135 points

Option

- Add Kübelwagen jeeps and 3-ton trucks for +5 points for the platoon.

Providing immediate heavy support for the panzergrenadiers were the 12cm sGW43 heavy mortars. Unlike gun artillery, these light weight weapons can be easily placed near the front lines with the panzergrenadeirs on defence or travel along with them and then quickly deploy on attack. It fires a heavy bomb giving it excellent destructive power, especially against dug-in enemy.

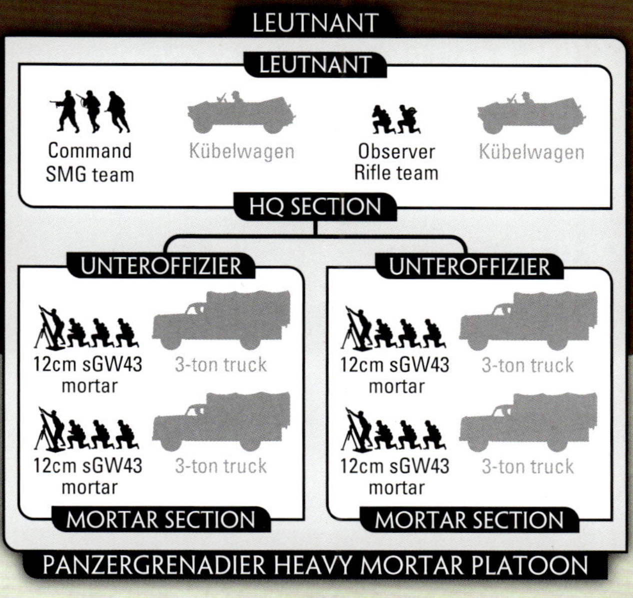

REGIMENTAL SUPPORT PLATOONS

Panzergrenadier Pioneer Platoon

Platoon

HQ Section with:

3 Pioneer Squads	200 points	185 points
2 Pioneer Squads	140 points	130 points

Options

- Add 3-ton trucks for +5 points for the platoon.
- Replace Command Pioneer Rifle/MG team with Command Pioneer Panzerfaust SMG team for +10 points.
- Equip one Pioneer Rifle/MG team with a Goliath demolition carrier in addition to its normal weapons for +30 points.
- Add Pioneer Supply 3-ton truck for +25 points or Pioneer Supply Maultier for +30 points.

You may replace up to one Pioneer Rifle/MG team per Pioneer Squad with a Flame-thrower team at the start of the game before deployment.

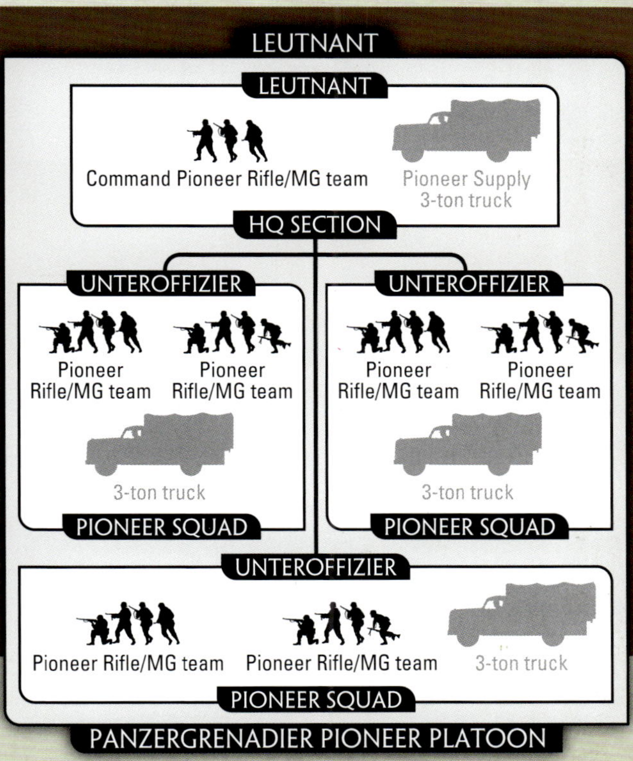

Panzergrenadier FlaK Platoon

Platoon		
6 Sd Kfz 251/21 D (Triple 15mm)	280 points	-
3 Sd Kfz 251/21 D (Triple 15mm)	140 points	-

Dual-purpose Sd Kfz 251/21 anti-aircraft half-tracks mount triple 15mm MG151 cannon originally produced for fighter aircraft. These weapons were used by *25. Panzergrenadierdivision* during Operation *Nordwind.*

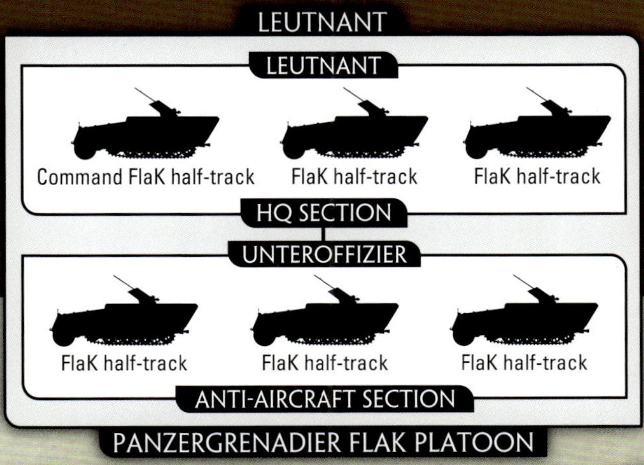

Panzergrenadier Infantry Gun Platoon

Platoon

HQ Section with:		
2 15cm sIG33	145 points	125 points

Option

- Add Kübelwagen jeep and Sd Kfz 11 half-tracks for +5 points for the platoon.

When the panzergrenadiers face particularly well dug-in enemy positions, they call up the heavy infantry guns. The 15cm sIG33 gun packs a heavy punch that will clear any enemy from their trenches or bunkers.

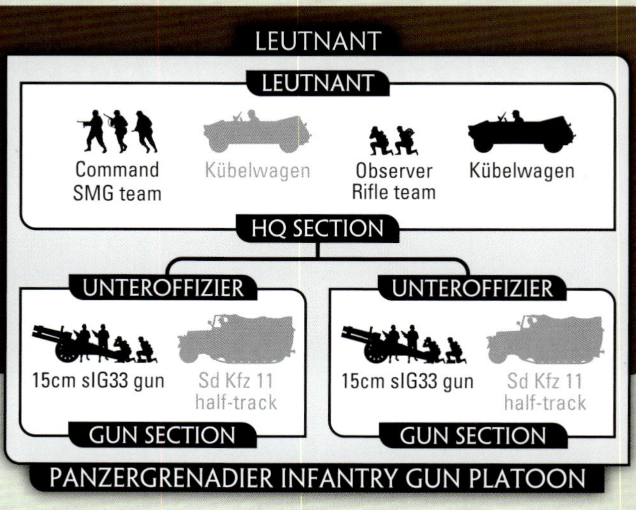

Armoured Panzergrenadier Platoon

Platoon

HQ Section with:		
3 Panzergrenadier Squads	285 points	-
2 Panzergrenadier Squads	200 points	-

Options

- Replace Command Panzerfaust MG team with Command Panzerfaust SMG team at no cost.
- Replace all Panzerfaust MG teams with Panzerfaust Assault Rifle teams for +5 points per team.

Armoured Panzergrenadier Platoons may use the Mounted Assault special rule (see page 243 of the rulebook).

119. Panzergrenadier Regiment's I Battalion (*25. Panzergrenadierdivision*) was mounted in Sd Kfz 251 half-tracks. They fought as part of *Kampfgruppe Huss* during the attacks on Hatten in Operation *Nordwind*.

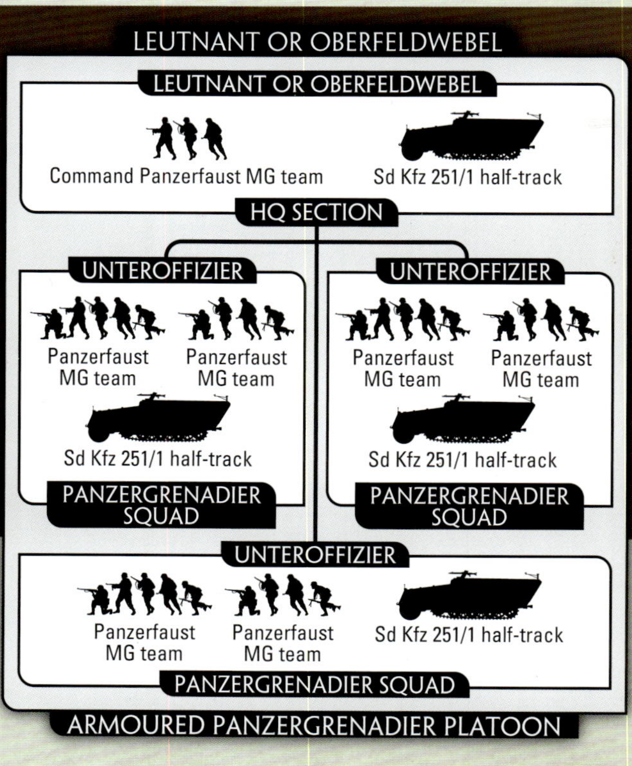

Support Platoons

Panzergrenadier Panzer Platoon

Platoon

3 StuG IV	-	250 points
2 StuG IV	-	165 points
3 StuG G (late)	-	260 points
3 StuG G (late)	-	175 points
3 Panther G	560 points	-
2 Panther G	375 points	-
3 Panzer IV/70 (A)	405 points	-
2 Panzer IV/70 (A)	270 points	-

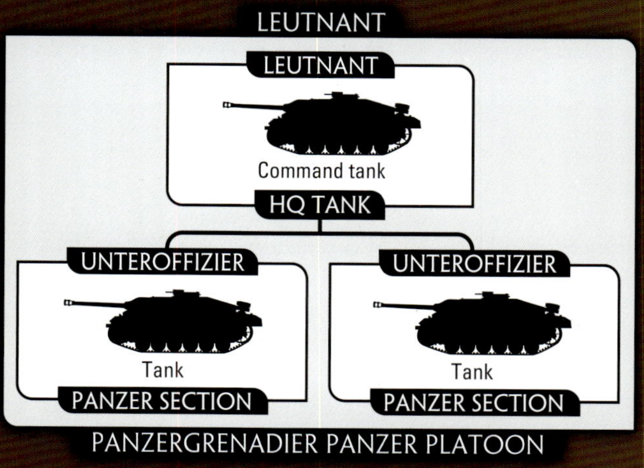

Flammpanzer Platoon

Platoon	Heer	SS
3 Flammpanzer Hetzer	170 points	170 points

A Panzergrenadierkompanie containing a Flammpanzer Platoon may use the Under the Cloak of Darkness special rule (see page 57).

A Flammpanzer Platoon is rated **Confident Veteran.**

CONFIDENT | VETERAN

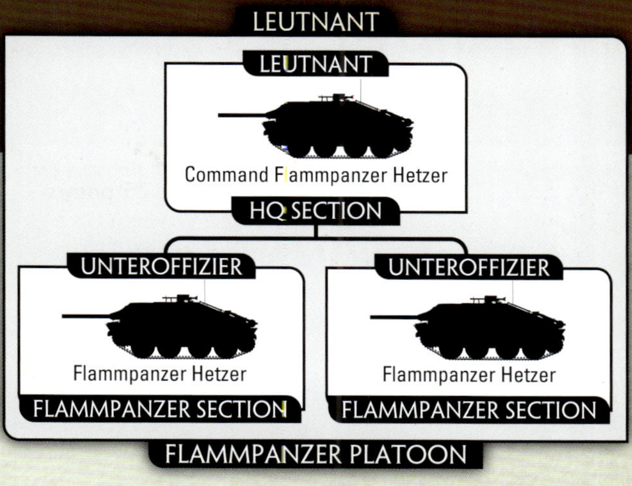

New *Flammpanzer 38(t) Hetzer* flame-tanks were rushed into service in time for Operation *Nordwind*. Flame-throwing tanks were ideal for taking out the tough bunkers of the Maginot Line being held by the Americans.

Two companies were ready, with *Flammpanzerkompanie 352* attached to *25. Panzergrenadierdivision* and *Flammpanzerkompanie 353* attached to *17. SS-Panzergrenadierdivision* for the operation.

On attack, the pioneers would cut through the wire and other obstacles and mark minefields under the cover of darkness. The *Flammpanzer 38(t) Hetzer* flame-tanks would then go into action, suppressing the bunkers with their flame projectors and allowing the pioneers to close on them with explosives to knock them out.

After breaking through the Maginot Line, *25. Panzergrenadierdivision* also later used their *Flammpanzer 38(t) Hetzer* flame-tanks in the street fighting for Rittershoffen.

Armoured Car Patrol

Platoon	Heer	SS
Sd Kfz 223 (radio) and 2 Sd Kfz 222 (2cm)	110 points	95 points
2 Sd Kfz 231 (8-rad)	80 points	70 points

Armoured Car Patrols are Reconnaissance Platoons.

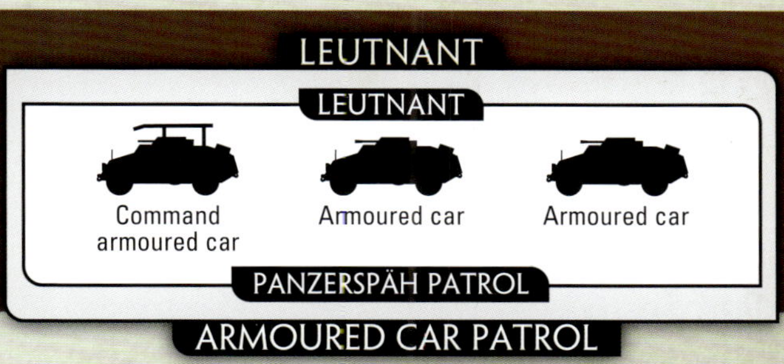

The *panzergrenadier* divisions had their own reconnaissance made up of older armoured cars like the Sd Kfz 222.

Panzergrenadier Anti-aircraft Gun Platoon

Platoon HQ Section with:	Heer	SS
3 3.7cm FlaK43	100 points	90 points
2 3.7cm FlaK43	70 points	60 points
3 2cm Flakvierling 38	95 points	80 points
2 2cm Flakvierling 38	65 points	55 points

Option

- Add 3-ton trucks for +5 points for the platoon.

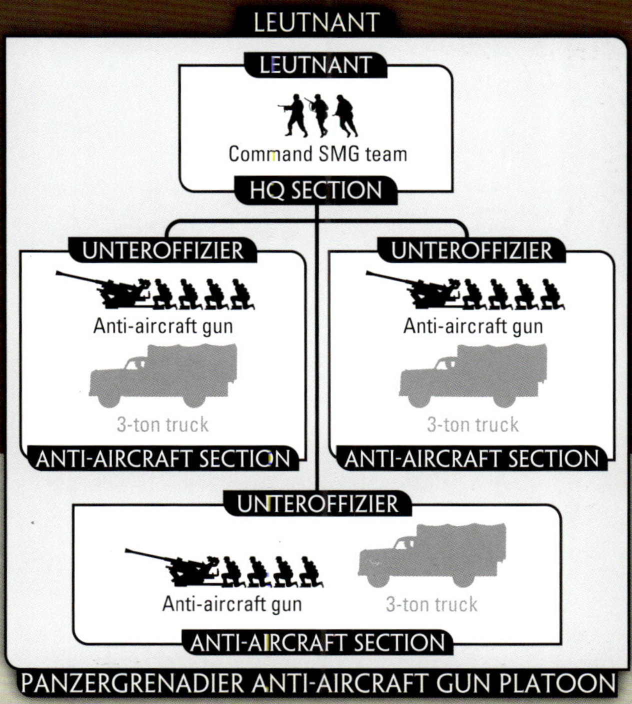

With the increasing number of attacks by Allied aircraft, the *panzergrenadier* anti-aircraft units have been issued with harder hitting and longer-ranged 3.7cm FlaK43 anti-aircraft guns. The four barrel 2cm Flakvierling gun has a lighter round, but a much higher rate of fire.

These new guns expand the umbrella of anti-air cover over the panzergrenadiers, giving them some protection from marauding P-47 fighter-bombers.

SCHWERE PANZERJÄGERKOMPANIE

HEAVY TANK-HUNTER COMPANY

TANK COMPANY

MOTIVATION AND SKILL

The vastly experienced men of the 653. Schwere Panzerjägerabteilung know how to use heavy tank-hunters. However, the early persistent breakdowns of the new vehicles has sapped the crews' initial confidence. A Schwere Panzerjägerkompanie is rated as **Reluctant Veteran.**

RELUCTANT	CONSCRIPT
CONFIDENT	TRAINED
FEARLESS	VETERAN

You must field one platoon from each box shaded black and may field one platoon from each box shaded grey.

HEADQUARTERS

HEADQUARTERS
- Schwere Panzerjäger-kompanie HQ — 67

COMBAT PLATOONS

ARMOUR
- Schwere Panzerjäger Platoon — 67

ARMOUR
- Schwere Panzerjäger Platoon — 67

ARMOUR
- Schwere Panzerjäger Platoon — 67

WEAPONS PLATOONS

ANTI-AIRCRAFT
- Panzerjäger Anti-aircraft Gun Platoon — 67

DIVISIONAL SUPPORT PLATOONS

INFANTRY
- Volksgrenadier Sturm Platoon — 47
- Volksgrenadier Schützen Platoon — 47
- Volks Pioneer Platoon — 51
- Panzergrenadier Platoon (SS) — 61
- Panzergrenadier Pionier Platoon (SS) — 64

INFANTRY
- Volksgrenadier Sturm Platoon — 47
- Panzergrenadier Platoon (SS) — 61

ARTILLERY
- Volks Light Artillery Battery — 52
- Volks Rocket Launcher Battery — 53
- Motorised Light Artillery Battery — 70
- Motorised Heavy Artillery Battery — 70

ANTI AIRCRAFT
- Panzergrenadier Anti-aircraft Gun Platoon (SS) — 65

AIRCRAFT
- Air Support — 68

HEADQUARTERS

Schwere Panzerjägerkompanie HQ

Headquarters

2 Jagdtiger	690 points
1 Jagdtiger	345 points

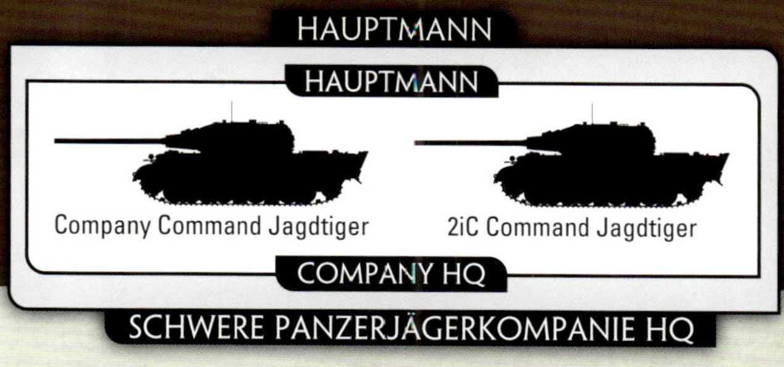

COMBAT PLATOONS

Schwere Panzerjäger Platoon

Platoon

4 Jagdtiger	1380 points
3 Jagdtiger	1035 points
2 Jagdtiger	690 points
1 Jagdtiger	345 points

The *Jagdtiger* heavy tank-hunter is an awesome beast to behold. This massive vehicle has a gun so powerful that no enemy tank in service can deflect its rounds. It has front armour so heavy that the enemy has no choice but to attempt to flank it to knock it out. All this power comes at a cost; its heavy weight makes it slow and mechanically unreliable.

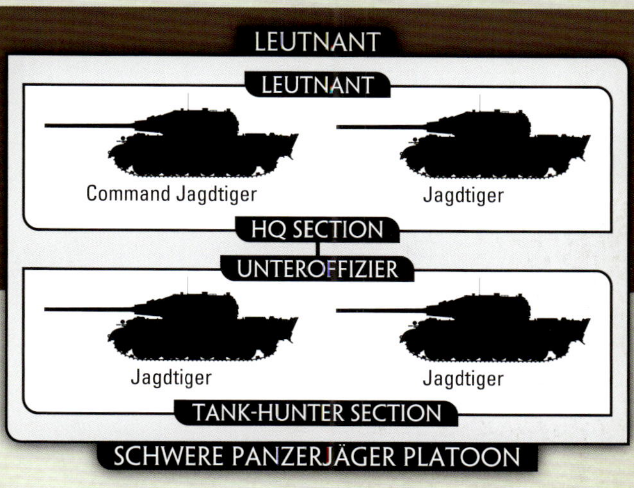

WEAPONS PLATOONS

Panzerjäger Anti-aircraft Gun Platoon

Platoon

4 Wirbelwind (Quad 2cm)	200 points
3 Wirbelwind (Quad 2cm)	150 points
2 Wirbelwind (Quad 2cm)	100 points
4 Möbelwagen (3.7cm)	200 points
3 Möbelwagen (3.7cm)	150 points
2 Möbelwagen (3.7cm)	100 points

With both *Wirbelwind* and *Möbelwagen* anti-aircraft tanks available, the *Jagdtiger* tank-hunters are well-protected from air attack.

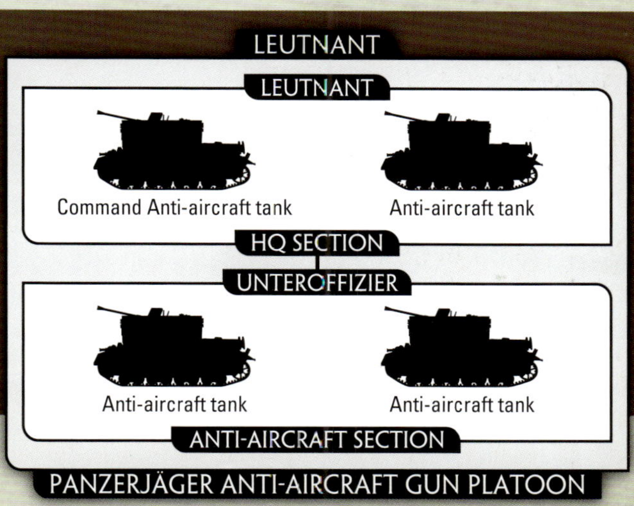

GERMAN DIVISIONAL SUPPORT

Motivation and Skill

25. Panzergrenadierdivision support platoons and support platoons from other Heer divisions are rated as **Confident Veteran** *unless otherwise noted.*

17. SS-Panzergrenadierdivision support platoons and support platoons from other SS divisions are rated as **Fearless Trained** *unless otherwise noted.*

HEER	
RELUCTANT	CONSCRIPT
CONFIDENT	TRAINED
FEARLESS	VETERAN

SS	
RELUCTANT	CONSCRIPT
CONFIDENT	TRAINED
FEARLESS	VETERAN

Air Support

Sporadic Air Support

Me 262 A2a Sturmvogel	105 points
FW 190F	100 points

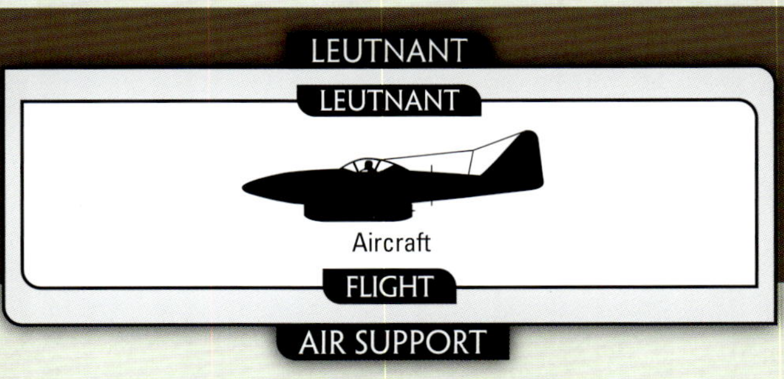

High-speed Jet

Ground attacks by Me 262 A2a aircraft cannot be intercepted using the Fighter Interception rule on page 179 of the rulebook.

The *Luftwaffe* (German Air Force) was not idle over the skies of Belgium, with fighter-bombers attacking targets of opportunity. However, it was also the debut of the world's first operational jet fighter: the Me 262. The ground attack version, the Me 262 A2a *Sturmvogel* (Storm Bird), was armed with two 30mm cannons and a pair of 250kg (550lb) bombs. Its excellent speed of 540mph (870km/h) meant that it could outrun any Allied fighter, making it near impossible to catch.

Operation Bodenplatte

Operation *Bodenplatte* (Baseplate) was to be launched on 16 December 1944. Its mission was to cripple Allied air support, giving the ground units of *Wacht Am Rhein* a fighting chance. However, the weather refused to co-operate and the operation wasn't launched until 1 January 1945. Over 900 aircraft took to the skies and struck the Allies' air bases in Belgium and Holland and destroyed somewhere between 300 and 400 Allied aircraft. This success, however, came at the steep price of over 300 German aircraft.

Heavy Anti-aircraft Gun Platoon

Platoon

	Heer	SS
2 AA Sections	165 points	140 points
1 AA Section	85 points	75 points

Option

- Model 8.8cm FlaK36 anti-aircraft guns with 8 or more crew and increase their ROF to 3 for +10 points per gun.

The anti-aircraft battalion of each *panzergrenadier* divisions contained two companies of the excellent 8.8cm FlaK36 anti-aircraft gun. The famous '88' was often used to smash enemy tanks as well as taking out aircraft.

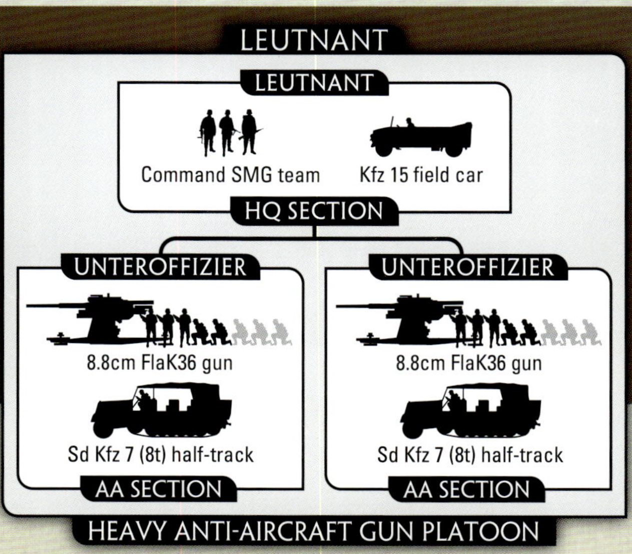

Panzer Platoon

Platoon		
5 Panzer IV J	-	375 points
4 Panzer IV J	350 points	300 points
3 Panzer IV J	260 points	225 points
5 Panther G	-	815 points
4 Panther G	750 points	655 points
3 Panther G	560 points	490 points

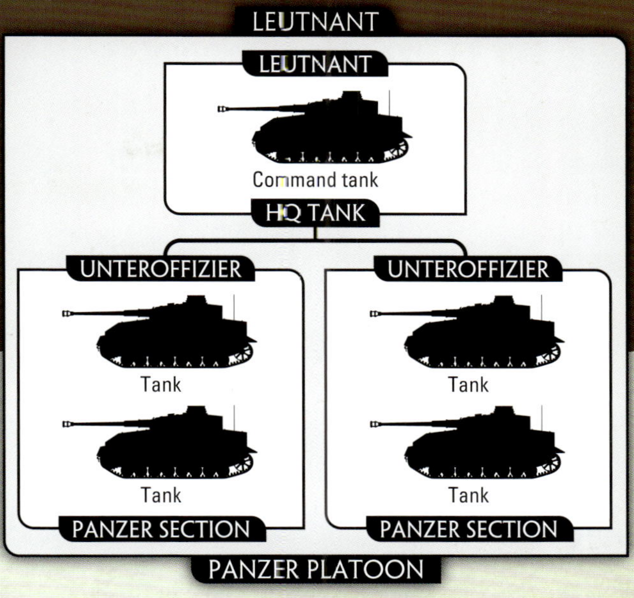

Volksgrenadier divisions were closely supported by panzer divisions during the Ardennes offensive. During the fighting for Bastogne, *26. Volksgrenadierdivision* was supported by tanks from *2. Panzerdivision*, *Panzer Lehr Division* and *15. Panzergrenadierdivision* at various times. Further north, *12. Volksgrenadierdivision* was supported by tanks from *12. SS-Panzerdivision* during a number of attacks.

Tank-hunter Platoon

Platoon		
3 StuG G or StuG IV	285 points	250 points
2 StuG G or StuG IV	190 points	165 points
3 StuG G (late)	295 points	260 points
2 StuG G (late)	195 points	175 points
3 Panzer IV/70 (V)	450 points	385 points
2 Panzer IV/70 (V)	300 points	255 points
4 Jagdpanther	920 points	-
3 Jagdpanther	690 points	-
2 Jagdpanther	460 points	-

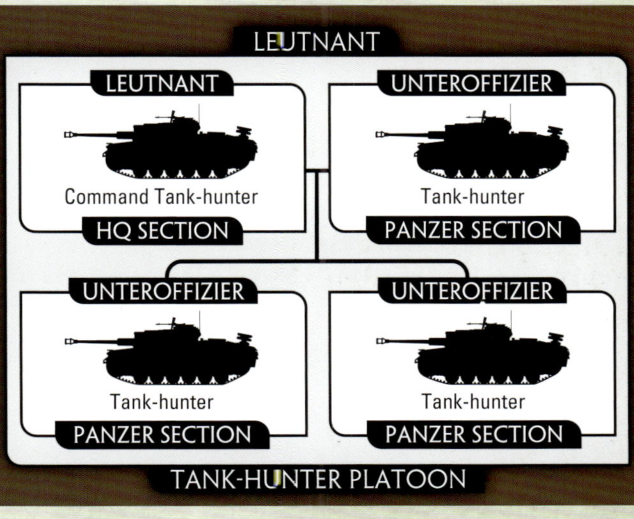

Sturmpanzer Platoon

Platoon	
4 Brummbär	280 points
3 Brummbär	210 points
2 Brummbär	140 points

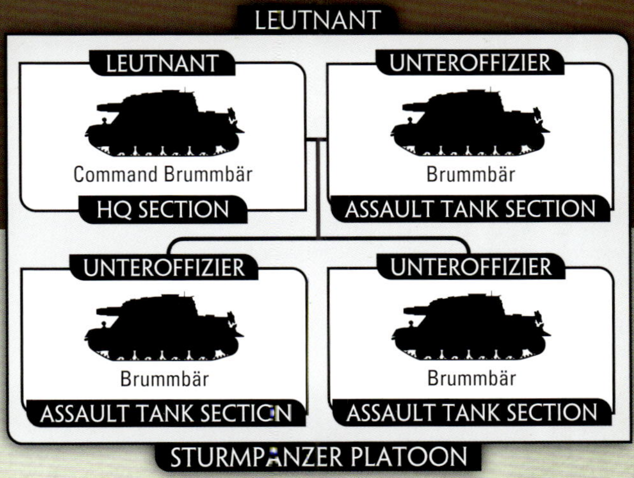

Due to the low number of operational StuG assault guns available to *12. Volksgrenadierdivision* for the Ardennes offensive, they were assigned *217. Sturmpanzer Abteilung* (217th Assault Tank Battalion) as additional support. The *Brummbär* (Grizzly Bear) assault tanks saw action with the Volksgrenadiers during the attacks on Dom Bütgenbach on 21 December 1944.

Motorised Light Artillery Battery

Platoon

HQ Section with:		
4 10.5cm leFH18/40	215 points	190 points
2 10.5cm leFH18/40	120 points	105 points

Option

- Add Kfz 68 radio truck, Kübelwagen jeeps and 3-ton trucks for +5 points for the battery.

You must purchase all the guns from one Gun Section before adding any teams from another Gun Section.

The artillery of the *panzergrenadier* divisions is motorised, making them better able to keep up with the rest of the division. This allows the battery to be in the right place at the right time to deliver its bombardments as they are needed.

The 10.5cm leFH18/40 howitzer is a lighter version that mounts the standard howitzer on a light weight modified PaK40 trail. To save the trail from shaking apart a large muzzle brake has been fitted to reduce its recoil.

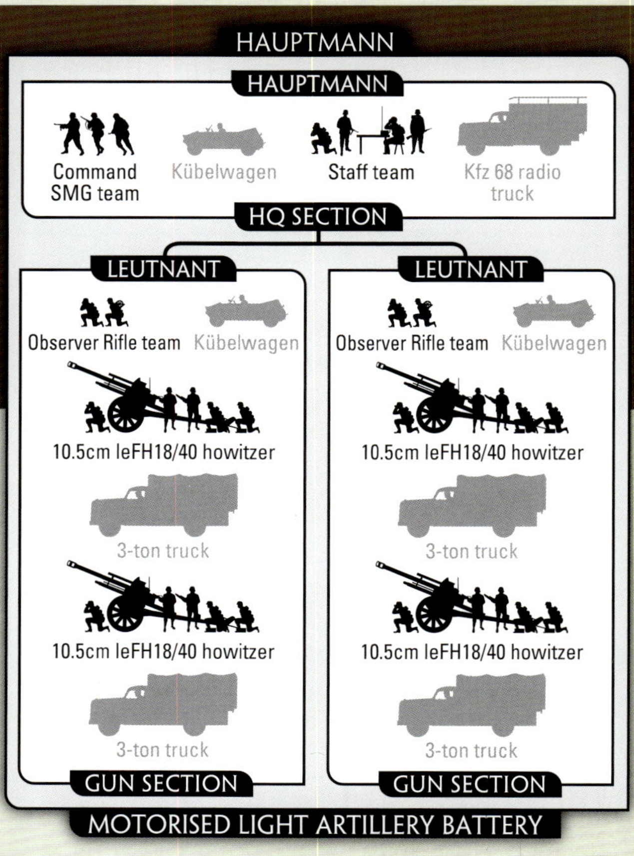

Motorised Heavy Artillery Battery

Platoon

HQ Section with:		
4 15cm sFH18	310 points	275 points
2 15cm sFH18	160 points	140 points
4 s10cm K18	395 points	340 points
2 s10cm K18	205 points	175 points

Option

- Add Kfz 68 radio truck, Kübelwagen jeeps and Kfz 7 half-tracks for +5 points for the battery.

You must purchase all the guns from one Gun Section before adding any teams from another Gun Section.

A *panzergrenadier* division's heavy firepower is supplied by the heavy artillery battalion armed with 15cm sFH18 howitzers and s10cm K18 guns.

The 15cm howitzer provided the hard hitting power to deal with enemy entrenchments and bunkers, while the long range 10cm guns are ideal for knocking out the enemy's artillery with counter battery fire.

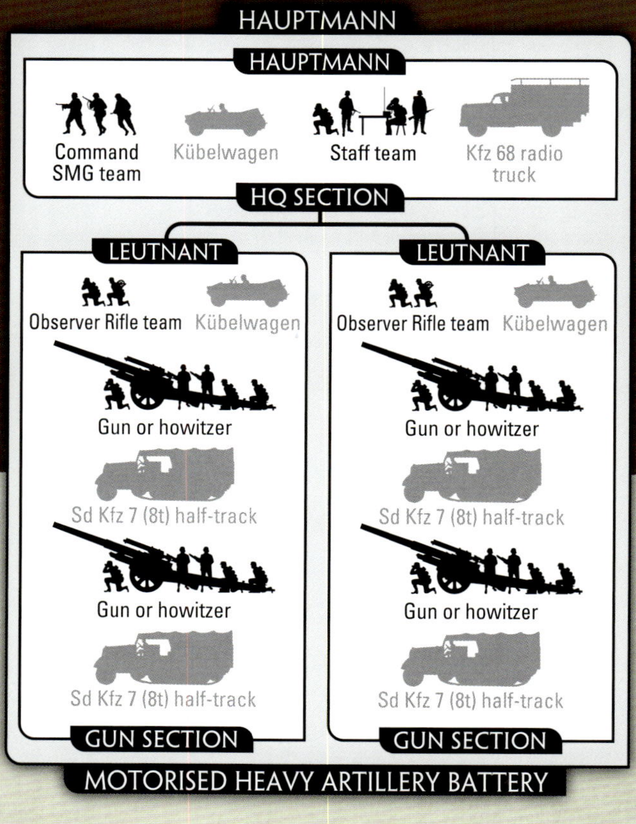

Heavy Assault Howitzer Platoon

Platoon

2 Sturmtiger	320 points
1 Sturmtiger	160 points

A Heavy Assault Howitzer Platoon is rated **Confident Trained.**

CONFIDENT | TRAINED

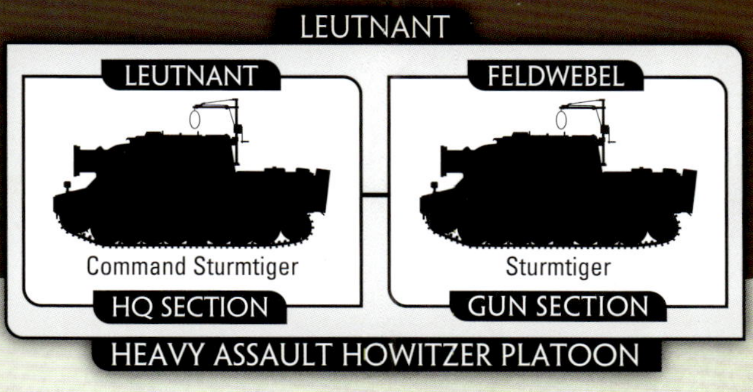

The *Sturmtiger* assault howitzers of *Sturmmörserkompanie 1000* and *1001* were attached to *6. SS-Panzer Armee* to take part in *Wacht am Rhein* (Watch on the Rhine, the Ardennes offensive) in mid-January 1944. However, it wasn't until late-January that they joined the battle where they were engaged in fighting on the Eifel front around Duren and Euskirchen. They were earmarked to lay waste to Liege once the offensive had broken through the American line.

Rocket Assault Howitzer

The 38cm RW61 assault howitzer of the *Sturmtiger* packs a powerful punch, but has a low ballistic speed when compared to a high velocity gun.

A 38cm RW61 assault howitzer has a minimum range of 8"/20cm.

Any Armoured vehicle hit by a 38cm RW61 assault howitzer takes an Armour Save using its Top Armour rating rather than its Front or Side Armour rating. Do not add an additional +1 to the target vehicle's top armour if the range is over 16"/40cm.

Any platoon hit by a 38cm RW61 assault howitzer is automatically Pinned Down.

Circumstances dictated, however, that most *Sturmtiger* engagements ended up being against Allied field forces rather than the bunkers and buildings they were designed for. These attacks made a significant impression on the few troops unfortunate enough to end up on the receiving end.

A Heavy Assault Howitzer Platoon may not launch Assaults and must Break Off rather than Counterattack.

Bunkerfeuer

Instead of using the normal shooting rules, a Sturmtiger may use the Bunkerfeuer rules when shooting its 38cm RW61 assault howitzer at Bunkers, teams in Buildings, or Gun teams.

To shoot at a Bunker, a team in a Building, or a Gun team, roll a Skill Test. It the test is successful the Sturmtiger has Hit the target.

Unlike other teams, a Sturmtiger can target Bunkers from ranges over 16"/40cm.

Shooting at Buildings

In addition to the normal Bunker Buster rules, if a 38cm RW61 assault howitzer Hits a team in a large building with multiple rooms, all the teams in rooms adjacent to the target team's room and those rooms above and below the target team's room are also Hit.

Sturmtiger (38cm RW61 auf Sturmmörser Tiger)

Hull MG

Front Armour 12

Movement: Slow Tank, Overloaded

Top Armour 2

Side Armour 8

Weapon	Range	ROF	AT	FP	Notes
38cm RW61 assault howitzer	48"/120cm	1	6	1+	Hull mounted, Awkward layout, Bunker buster, Bunkerfeuer, Rocket assault howitzer, Minimum range 8"/20cm.

Crew:	5
Battle weight:	65 tonnes
Length:	6.28m
Width:	3.57m
Height:	2.85m
Engine:	650 HP
Main gun:	380mm
Ammunition Carried:	14 rounds
Armour:	25-150mm
Max speed:	40km/h
Range on road:	120km
Total production:	18

FIELDING SKORZENY COMMANDO GROUPS

The role of Skorzeny's Commandos was to sew the seeds of panic and confusion amongst the ranks of the American defenders during the opening stages of the Ardennes offensive. This would allow the hard-hitting panzers to then smash through the line of the disorganised enemy.

Motivation and Skill

Skorzeny Commando teams are rated **Fearless Veteran.**

RELUCTANT	CONSCRIPT
CONFIDENT	TRAINED
FEARLESS	VETERAN

Skorzeny Commando Group

2 Skorzeny Commando teams	100 points
1 Skorzeny Commando team	50 points

SKORZENY COMMANDO GROUPS IN FLAMES OF WAR

Attributes

Skorzeny Commando teams are unarmed Independent Jeep teams. Being composed entirely of Independent teams, a Skorzeny Commando Group is not a platoon for any purpose whatsoever.

Perfect Disguises

Skorzeny Commando teams are considered friendly teams by your opponent's force, although they do not consider the opponent's teams to be friendly teams. As such, they can never be shot at, used as the Aiming Point for an Artillery Bombardment, or assaulted by enemy teams. Nor can Skorzeny Commando teams shoot, launch an assault, or perform a Stormtroopers move. However, unlike normal friendly teams, enemy teams can shoot through Skorzeny Commando teams.

Ignore Skorzeny Commando teams completely during artillery bombardments or air attacks and when determining the Safety Distance for aircraft.

Deployment

Skorzeny Commando teams are deployed in your opponent's Deployment Area, immediately after your opponent's Independent Teams have been deployed.

Movement

Skorzeny Commando teams may only move during your opponent's Movement Step after all of your opponent's teams have moved. Skorzeny Commando teams can move within 2"/5cm of enemy teams, but can never move At the Double.

Time to Head Back

If a Skorzeny Commando team ends its Movement Step within 16"/40cm of a friendly team (other than another Skorzeny Commando team) it is removed from play.

Exposed

If a Skorzeny Commando team is Exposed (either by being Challenged or by failing in an attempt to Sow Chaos), it is Destroyed and immediately removed from play. This is the only way a Skorzeny Commando team can be Destroyed.

Challenge

During your opponent's Shooting Step, your opponent may Challenge a Skorzeny Commando team with any of their Command teams within 6"/15cm. Instead of shooting, each command team that wishes to, may immediately Challenge the Skorzeny Commando team (the rest of the platoon may shoot as normal).

Each time it is Challenged, the Skorzeny Commando team must make an immediate Skill Check:

- If successful, the Skorzeny Commando team answers the challenge correctly and is not Exposed.
- If unsuccessful, the Skorzeny Commando team fails to give the correct answer and is Exposed.

SOWING CHAOS BEHIND ENEMY LINES

The activities of *Unternehmen Greif* (Operation Griffin) were legendary, far beyond truth in many cases! For their size, the Skorzeny Commandos sowed a extraordinary amount of chaos in the early days of the Ardennes offensive. Even Patton's troops as far south as the Lorraine were convinced Skorzeny's men were operating in their area!

During your opponent's Starting Step, Skorzeny Commando teams may attempt to sow chaos amongst your opponent's troops. Once per turn, each Skorzeny Commando team may choose one of the following special rules to use.

Spread Rumours

Perhaps the best result of *Unternehmen Greif* was the impact of the false rumours and contradictory orders spread by the commandos. The most effective rumour spread by Skorzeny's men was that their ultimate mission was to infiltrate as far as Paris to assassinate General Eisenhower, overall commander of Allied forces. The general's staff was so paranoid they locked Eisenhower away, making it difficult for him to conduct the war!

At the beginning of your opponent's Starting Step, immediately before the opponent Rallies Pinned Down Platoons, this Skorzeny Commando team may attempt to Spread Rumours. Choose an enemy platoon within 4"/10cm of this Skorzeny Commando team and Roll a Skill Check:

- *If successful, the commandos spread devious rumours amongst the men of the enemy platoon and it becomes Pinned Down.*
- *Otherwise, the rumours raise suspicion and the Skorzeny Commando team is immediately Exposed.*

Reverse Road Signs

One easy way to disrupt an enemy counterattack is to switch road signs or mark off safe roads with white tape to convince the enemy the road is mined.

At the beginning of your opponent's Starting Step, immediately before the opponent roll for Reserves, this Skorzeny Commando team may attempt to Reverse Road Signs. Roll a Skill Check for the Skorzeny Commando team:

- *If successful, your opponent's Reserves may not arrive anywhere within 12"/30cm of this Skorzeny Commando team during this turn.*
- *Otherwise, the Skorzeny Commando team is caught in the act and it is Exposed.*

If a platoon cannot arrive on the table as a result of this special rule, it returns to Reserve instead.

Observe Defences

The commando teams collected information on enemy defences, troop strength, and other details that would be useful for the advancing German Army.

At the beginning of your opponent's Starting Step, immediately before the opponent Reveals Ambushes, this Skorzeny Commando team may Observe Defences. Roll a Skill Check for the Skorzeny Commando team:

- *If successful, your opponent cannot place teams from Ambush within 8"/20cm and Line Of Sight of this Skorzeny Commando team.*
- *Otherwise, the Skorzeny Commando team draws too much attention to itself and it is Exposed.*

GERMAN ARSENAL

TANK TEAMS

Name *Weapon*	Mobility *Range*	Armour Front *ROF*	Side *Anti-tank*	Top *Firepower*	Equipment and Notes
TANKS					
Panzer IV J *7.5cm KwK40 gun*	Standard Tank *32"/80cm*	6 *2*	3 *11*	1 *3+*	Co-ax MG, Hull MG, Protected ammo, Schürzen. *Slow traverse.*
Panther G *7.5cm KwK42 gun*	Standard Tank *32"/80cm*	10 *2*	5 *14*	1 *3+*	Co-ax MG, Hull MG, Wide tracks.
FLAME-THROWER TANKS					
Flammpanzer Hetzer *1.4cm Flammenwerfer*	Standard Tank *4"/10cm*	7 *3*	2 *-*	1 *5+*	Hull MG, Overloaded. *Hull mounted, Flame-thrower.*
TANK HUNTERS					
Jagdtiger *12.8cm PaK44 gun*	Slow Tank *48"/120cm*	16 *2*	8 *17*	2 *2+*	Hull MG, Overloaded, Unreliable. *Hull mounted, Breakthrough gun.*
Jagdpanther *8.8cm PaK43 gun*	Standard Tank *40"/100cm*	10 *2*	5 *16*	1 *3+*	Hull MG. *Hull mounted.*
Panzer IV/70 (A) *7.5cm PaK42 gun*	Slow Tank *32"/80cm*	8 *2*	3 *14*	1 *3+*	Hull MG, Overloaded, Schürzen. *Hull mounted.*
Panzer IV/70 (V) *7.5cm PaK42 gun*	Slow Tank *32"/80cm*	9 *2*	3 *14*	1 *3+*	Hull MG, Overloaded, Schürzen. *Hull mounted.*
Hetzer *7.5cm PaK39 gun*	Standard Tank *32"/80cm*	7 *2*	2 *11*	1 *3+*	Hull MG, Overloaded. *Hull mounted.*
ASSAULT GUNS					
StuG G or StuG IV *7.5cm StuK40 gun*	Standard Tank *32"/80cm*	7 *2*	3 *11*	1 *3+*	Hull MG, Protected ammo, Schürzen. *Hull mounted.*
StuG G (late) *7.5cm StuK40 gun*	Standard Tank *32"/80cm*	7 *2*	3 *11*	1 *3+*	Co-ax MG, Hull MG, Protected ammo, Schürzen. *Hull mounted.*
Brummbär *15cm StuH43 gun*	Slow Tank *16"/40cm*	9 *1*	5 *13*	1 *1+*	Hull MG, Overloaded, Schürzen. *Hull mounted, Bunker buster.*
Sturmtiger *38cm RW61 assault howitzer*	Slow Tank *48"/120cm*	12 *1*	8 *6*	2 *1+*	Hull MG, Overloaded. *Hull mounted, Awkward layout, Bunker buster, Bunkerfeuer, Rocket assault howitzer, Minimum range 8"/20cm.*
ANTI-AIRCRAFT					
Möbelwagen (3.7cm) *3.7cm FlaK43 gun*	Standard Tank *24"/60cm*	0 *4*	0 *6*	0 *4+*	*Anti-aircraft.*
Wirbelwind (Quad 2cm) *2cm FlaK38 (V) gun*	Standard Tank *16"/40cm*	3 *6*	1 *5*	0 *5+*	Hull MG. *Anti-aircraft.*
Sd Kfz 251/21 D (Triple 15mm) *MG151/15 gun*	Half-tracked *16"/40cm*	1 *6*	0 *4*	0 *5+*	*Anti-aircraft.*
RECONNAISSANCE					
Sd Kfz 222 (2cm) *2cm KwK38 gun*	Wheeled *16"/40cm*	1 *3*	0 *5*	0 *5+*	Co-ax MG. *Self-defence anti-aircraft.*
Sd Kfz 223 (radio)	Wheeled	1	0	0	AA MG.
Sd Kfz 231 (8-rad) *2cm KwK38 gun*	Jeep *16"/40cm*	2 *3*	0 *5*	0 *5+*	Co-ax MG.
VEHICLE MACHINE-GUNS					
Vehicle MG	*16"/40cm*	*3*	*2*	*6*	*ROF 1 if other weapons fire.*

GUN TEAMS

Weapon	Mobility	Range	ROF	Anti-tank	Firepower	Notes
MG42 HMG	Man-packed	24"/60cm	6	2	6	ROF 3 when pinned down or moving.
8cm GW34 mortar	Man-packed	24"/60cm	2	2	3+	Smoke, Minimum range 8"/20cm.
Firing bombardments		40"/100cm	-	2	6	Smoke bombardment.
12cm sGW43 mortar	Light	56"/140cm	-	3	3+	
7.5cm IG37 gun	Light	16"/40cm	2	9	3+	Gun shield, Smoke.
Firing bombardments		56"/140cm	-	3	6	
7.5cm PaK50 gun	Medium	24"/60cm	2	10	3+	Gun shield.
Firing bombardments		64"/160cm	-	3	6	
15cm sIG33 gun	Heavy	16"/40cm	1	13	1+	Bunker buster, Gun shield.
Firing bombardments		56"/140cm	-	4	2+	
2cm Flakvierling 38 gun	Immobile	16"/40cm	6	5	5+	Gun shield, Anti-aircraft, Turntable.
3.7cm FlaK43 gun	Immobile	24"/60cm	4	6	4+	Gun shield, Anti-aircraft, Turntable.
8.8cm FlaK36 gun	Immobile	40"/100cm	2	13	3+	Gun shield, Heavy anti-aircraft, Turntable.
7.62cm PaK36(r) gun	Heavy	32"/80cm	2	11	3+	Gun shield.
7.5cm PaK40 gun	Medium	32"/80cm	2	12	3+	Gun shield.
7.5cm FK40 (PaK40) gun	Medium	32"/80cm	2	12	3+	Gun shield.
Firing bombardments		64"/160cm	-	3	6	
8.8cm FK43 (PaK43) gun	Immobile	40"/100cm	2	16	3+	Gun shield, Turntable.
Firing bombardments		88"/220cm	-	3	5+	
s10cm K18 gun	Immobile	32"/80cm	1	15	2+	
Firing bombardments		96"/240cm	-	4	4+	
10.5cm leFH18/40 gun	Heavy	24"/60cm	1	10	2+	Breakthrough gun, Gun shield, Smoke.
Firing bombardments		72"/180cm	-	4	4+	Smoke bombardment.
15cm sFH18 gun	Immobile	24"/60cm	1	13	1+	Bunker buster, Smoke.
Firing bombardments		80"/200cm	-	5	2+	Smoke bombardment.
15.2cm KH433(r) (ML-20) gun	Immobile	32"/80cm	1	13	1+	Bunker buster, Gun shield.
Firing bombardments		88"/220cm	-	5	2+	
15cm NW41 rocket launcher	Light	64"/160cm	-	3	4+	Rocket launcher, Smoke bombardment.
21cm NW42 rocket launcher	Light	72"/180cm	-	3	3+	Rocket launcher.
30cm NW42 rocket launcher	Light	56"/140cm	-	3	1+	Rocket launcher, Super-heavy rockets.

INFANTRY TEAMS

Team	Range	ROF	Anti-tank	Firepower	Notes
Rifle team	16"/40cm	1	2	6	
Rifle/MG team	16"/40cm	2	2	6	
MG team	16"/40cm	3	2	6	ROF 2 when pinned down.
SMG team	4"/10cm	3	1	6	Full ROF when moving.
Assault Rifle team	8"/20cm	3	1	6	Full ROF when moving.
Flame-thrower team	4"/10cm	2	-	6	Flame-thrower.
Panzerschreck team	8"/20cm	2	11	5+	Tank assault 5.
Staff team	16"/40cm	1	2	6	Moves as a Heavy Gun team.
ADDITIONAL TRAINING AND EQUIPMENT					
Panzerfaust	4"/10cm	1	12	5+	Tank Assault 6, Cannot shoot in the Shooting Step if moved in the Movement Step.

Pioneer teams are rated as Tank Assault 4.

TRANSPORT TEAMS

Vehicle *Weapon*	Mobility *Range*	Armour Front *ROF*	 Side *Anti-tank*	 Top *Firepower*	Equipment and Notes
TRUCKS					
Kübelwagen jeep or captured Jeep	Jeep	-	-	-	
Steyr Kfz 70, Opel Blitz 3-ton, or Kfz 68 Radio truck	Wheeled	-	-	-	
Opel Maultier	Half-tracked	-	-	-	
Horse-drawn wagon	Wagon	-	-	-	
TRACTORS					
Sd Kfz 7 or Sd Kfz 11 half-track	Half-tracked	-	-	-	
RSO	Slow Tank	-	-	-	
Horse-drawn limber	Wagon	-	-	-	
ARMOURED PERSONNEL CARRIERS					
Sd Kfz 251/1 D half-track	Half-tracked	1	0	0	Hull MG, Passenger-fired AA MG.

AIRCRAFT

Aircraft	Weapon	To Hit	Anti-tank	Firepower	Notes
Me 262 A2a Sturmvogel	Cannon Bombs	3+ 4+	9 5	5+ 2+	High-speed Jet.
FW 190F	Cannon Bombs	3+ 4+	7 5	5+ 1+	

PAINTING GERMANS

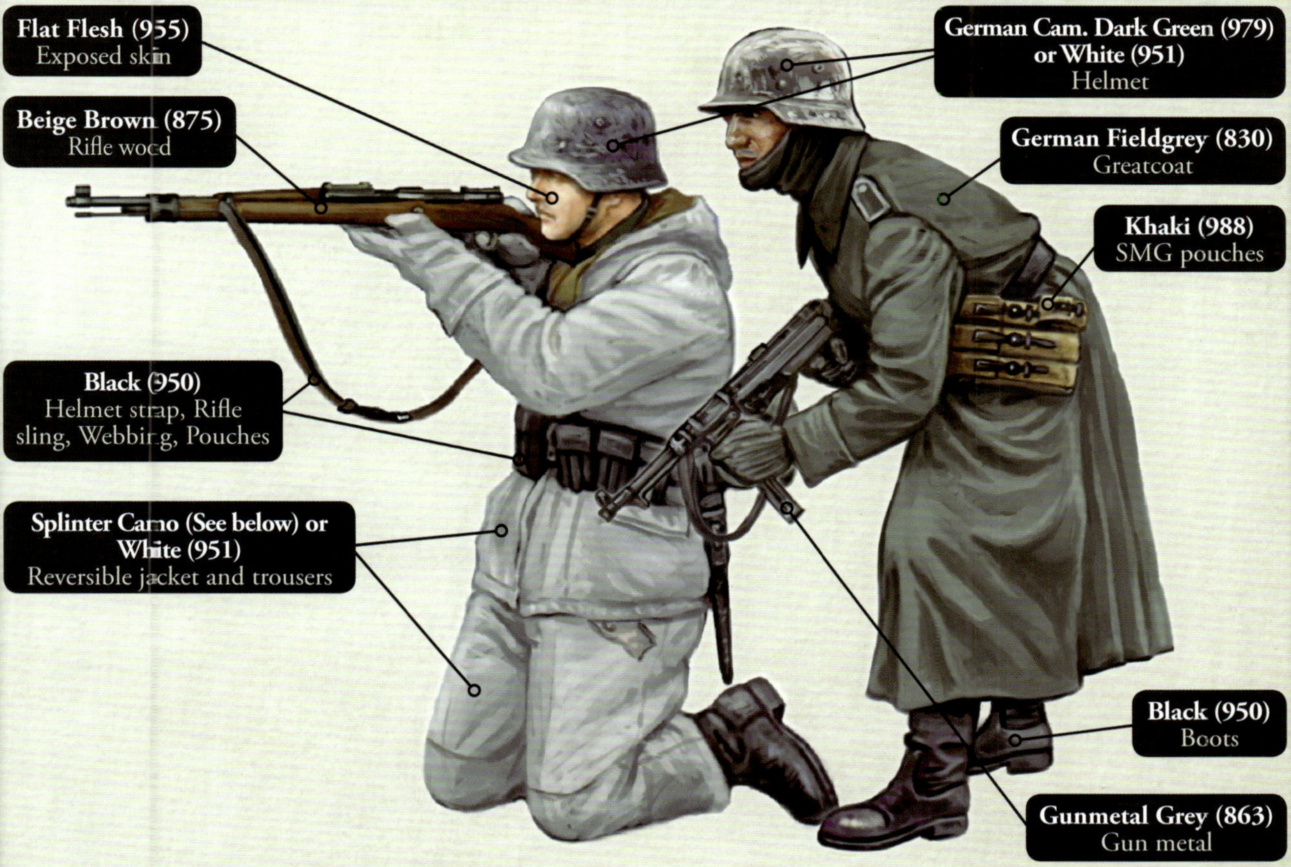

SPLINTER PATTERN CAMOUFLAGE

1. Start with a base of **German Camo Beige (821)**.
2. Paint irregular lines and patches of **German Camo Medium Brown (826)**. Try to create angular, zig-zag shapes; you should aim to cover approximately one-third of the base colour, at most.
3. Add small patches of **Luftwaffe Camo Green (823)** or **Flat Green (968)** between the brown areas.

This basic method ignores the vertical lines – Splinters – that give this camo pattern its name. Some people like to paint fine lines, but a more scale-appropriate method is to use a base colour which suggests the Splinters – **Green Grey (886)** works well – using small patches of **German Camo Beige (821)** to indicate the areas without Splinters.

PAINTING WHITE

Apply a base coat of a mid grey such as **Medium Sea Grey (991)**. This shadow colour doesn't need to be perfectly smooth, but try to make the edges neat.

Using a pale grey such as **Sky Grey (989)**, cover most of the base coat, leaving a small amount of the darker grey showing in the deeper recesses.

Paint pure **White (951)** highlights on raised areas. Pale colours lose intensity as they dry, so don't worry if your highlights look a bit garish at first.

Pale colours such as white can be a challenge to paint neatly, But with a little extra effort and patience, you will be rewarded with bright, eye-catching figures.

ARDENNES BATTLEFIELDS

The Ardennes is a region of hills, rivers, and woods covering southeast Belgium, the neighbouring area of France, and the small state of Luxembourg. When viewed with the tactical eye, the Ardennes seems an unlikely place to drive a major offensive through, an assumption that the Germans used successfully against their enemies in both 1914 and 1940. In late 1944 they once again planned to push through the Ardennes into the heart of Belgium through its picturesque backdoor.

Around a third of the Ardennes is covered with forest, much of which is coniferous plantations with neat rows of trees. The forests are scattered in a patchwork across the countryside. The landscape is cut through with winding rivers and streams.

The road network in 1944 was quite extensive, despite the low population and industry of the region. The roads were expanded and improved in the 1930s with the growth of automobile tourism. Not a single main highway traversing the Ardennes was in a straight east-west direction, making the intersections and crossroads important for traversing the region. The area also is dotted with towns and villages.

VILLAGES

THE ARDENNES REGION IS DOTTED WITH SMALL VILLAGES AND FARMSTEADS. AN AVERAGE VILLAGE IN THE ARDENNES CONSISTED OF STONE HOUSES AND VERY NARROW, WINDING STREETS. THESE VILLAGES OFTEN CONSTRICTED THE THROUGH ROAD TO SINGLE-LANE TRAFFIC. LONE FARMSTEADS OR INNS OFTEN GAVE THEIR NAMES TO THE CROSSROADS ON WHICH THEY STAND.

OPEN GROUND

BETWEEN THE FORESTS, THE ARDENNES IS A PATCHWORK OF FIELDS AND PASTURE WHICH IS CROSS-COUNTRY TERRAIN. PASTURES ARE OFTEN DIVIDED BY BARBED WIRE FENCES, THESE ARE SLOW GOING LINEAR OBSTACLES.

ROADS & TRACKS

ALL OF THE MAIN ROADS OF THE ARDENNES HAVE HARD SURFACES, ALTHOUGH MOST OF THE MINOR ROADS ARE UNPAVED. ON SHADED ROADS, THAT RUN THROUGH WOODS, YOU MAY LIKE TO USE THE ICE RULES ON PAGE 80 TO REPRESENT ICE THAT HASN'T THAWED DURING THE DAY.

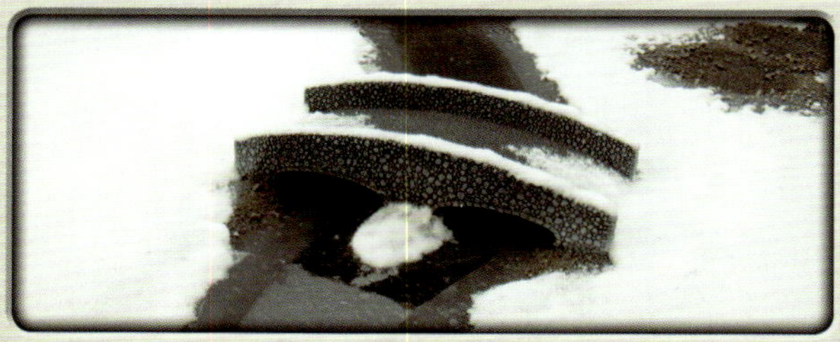

STREAMS & RIVERS

THE ARDENNES IS CUT THROUGH WITH WINDING RIVERS AND STREAMS, SOME FLOWING PARALLEL TO RIDGES, WHILE OTHERS CUT THROUGH HIGHER GROUND. STREAMS ARE DIFFICULT GOING WITH SLOW GOING FORDS WHERE TRACKS CROSS. RIVERS ARE WIDER THAN STREAMS AND ARE VERY DIFFICULT GOING. FORDS ACROSS RIVERS ARE DIFFICULT GOING, BUT MOST ROADS CROSS BY BRIDGES.

WOODS

MANY ARDENNES FORESTS HAVE BEEN REPLANTED IN MASSIVE GROVES; THE TREES PLANTED IN LONG AND PRECISE ROWS WITH LARGE GAPS IN BETWEEN. THE SANDY SOIL PREVENTS UNDERGROWTH, SO IT IS EASY TO SEE LONG DISTANCES INTO THE WOODS.

YOU CAN MODEL THIS ON YOUR BATTLEFIELD BY COVERING A MAJORITY OF YOUR TABLE WITH WOODED AREAS. RATHER THAN TREATING THESE FORESTS AS AREA TERRAIN AS USUAL, SIMPLY MAKE THE AREA THEY OCCUPY CONCEAL THOSE TEAMS INSIDE. TO REPRESENT THE COMPLICATIONS OF MANOEUVRING VEHICLES IN FORESTS, THEY SHOULD BE RATED AS SLOW GOING (RATHER THAN DIFFICULT OR VERY DIFFICULT GOING AS USUAL).

ARTILLERY DAMAGED WOODS

AS THE FIGHTING AROUND BASTOGNE INTENSIFIED MANY ARTILLERY BARRAGES WERE DROPPED ON AMERICAN AND GERMAN POSITIONS IN WOODS. A BARRAGE WOULD LEAVE BROKEN AND FALLEN TREES AND CRATERED GROUND IN ITS WAKE. THESE ARTILLERY SMASHED WOODS ARE DIFFICULT GOING.

SNOW

ONCE THE SNOW BEGAN TO FALL AROUND BASTOGNE IT ADDED ONE MORE DIFFICULTY TO THE FIGHTING. THE SNOW COVERS THE GROUND, MAKING IT HARD TO LOCATE STREAMS, FENCES AND UNEVEN GROUND. THIS MEANS TROOPS HAVE TO BE MORE CAUTIOUS IN SNOW COVERED TERRAIN. TO PLAY A GAME IN THE SNOW, USE THE SNOW-COVERED TERRAIN RULES ON PAGE 80.

BRIDGE DEMOLITION

All across the Ardennes front the American engineers worked hard to destroy the very bridges they had rebuilt only a few weeks before. The destruction of these bridges slowed down the advance of the Germans.

Agree with your opponent which bridges have been prepared for demolition.

If at the beginning of their Movement Step, a player has one or more Pioneer teams adjacent to a bridge that has been prepared for demolition, roll a die:

- *If the result is 6, the bridge is demolished and collapses. Any teams on the bridge are immediately Destroyed.*
- *Otherwise the bridge was not ready for demolition and nothing happens this turn.*

A demolished bridge becomes Impassable for vehicles and Gun teams, however it is only Very Difficult Going for Infantry and Man-packed-gun teams.

RICKETY BRIDGES

There was a broad assortment of bridges that spanned the many rivers in the Ardennes. Some were sturdy stone structures, while others were much lighter and never meant to support the weight of a tank!

Agree with your opponent which bridges are rickety. Typically, a rickety bridge is one that is made of wood, as opposed to one made from stone or metal. However, even ones that look sturdy can be structurally unsound due to artillery or botched demolition attempts.

When an Armoured vehicle attempts to cross a Rickety Bridge, it must roll a die and add 4 to the result.

- *If the total is less than the vehicle's Front armour, the bridge collapses under its weight and the vehicle is Destroyed.*
- *Otherwise the vehicle safely navigates the bridge and it may continue as normal.*

A collapsed bridge becomes Impassable for vehicles and Gun teams, however it is only Very Difficult Going for Infantry and Man-packed-gun teams.

WINTER WEATHER

Perhaps the single most defining aspect of the Ardennes campaign was the weather. Certainly, the Germans knew that as they carefully planned the offensive for the time when the weather would ground Allied planes, allowing their panzers to travel in the daylight.

However, the bad weather was a double-edged sword as the freezing rain mixed with the mud to create terrible conditions on the roads. Deep snow piled up in the few clearings, making it tough for troops to slog their way through. The white snow blinded the troops as they kept a vigilant watch from their freezing foxholes.

What follows are several optional rules for representing bad weather in your games.

Your Battlefield

These rules are optional. Feel free to use all of them, but be aware that this can lead to slower and longer games.

If that isn't your style but you'd still like to add some flavour to your battle, go ahead and choose a few of the rules to use in your game and leave the others for another time and another battle.

OPTIONAL WINTER BATTLE SPECIAL RULES

Snow-covered Terrain

Moving in the snow presents a number of hazards for the unwary traveller. Deep snow impedes foot traffic as much as it does vehicle movement.

All teams treat snow-covered Roads as Cross-country Terrain, and all other snow-covered Terrain as Slow Going.

Ice

Packed snow and ice make moving vehicles a risky affair. Black ice, which is extremely hard to spot, makes honest men out of even the most reckless drivers!

If a vehicle moves more than 8"/20cm in the Movement Step, at the end of its movement roll a Skill Check for that vehicle.

- *If successful, the vehicle navigates through the ice and carries on unharmed.*
- *Otherwise, the vehicle slips on the ice and becomes Bogged Down.*

Frozen Rivers and Ponds

Frozen rivers, streams, lakes, and ponds posed their own threats, not least of which was the temptation to cross them with armoured vehicles to bypass the destroyed bridges.

For light vehicles, such as trucks, jeeps, and other utility vehicles, a frozen river is easier to cross. For these vehicles a frozen river is rated as Cross-country terrain covered in Ice (see above).

However, a crossing is far more risky for armoured vehicles. When an Armoured vehicle attempts to cross a frozen river, it must roll a die and add 3 to the result.

- *If the total is less than the vehicle's Front armour, the ice gives way under the vehicle's weight and the vehicle is Destroyed.*
- *Otherwise the vehicle safely navigates the river and it may continue as normal.*

Snow Storm

Snow storms were frequent in January 1945, and caused confusion and frustration.

If both players agree, your battle can take place during a snow storm. The game is played under the Night Fighting rules in the rulebook (see page 272 of the rulebook). The mission is played at Dawn (see page 273 of the rulebook).

Blizzard

Blizzards are a more serious breed of snow storm. They can strike suddenly, delivering a pile of snow within a matter of minutes.

If both players agree you can decide to risk having a blizzard strike your battlefield.

If a Blizzard could strike in your battle, roll a die at the start of each player's turn. On a 1, the blizzard strikes. That turn and those following are played under the Night Fighting rules in the rulebook (pages 272 and 273).

Instead of using the Night Visibility Table, use the Blizzard Visibility Table below to find out how far your teams can see in the Blizzard.

Players continue to roll at the start of their turn. On a 5+ the Blizzard dies off, and the rest of the game uses the normal visibility rules.

If the game is not already using the Snow-covered Terrain special rule when the Blizzard strikes, the rest of the game is played with them in effect.

Blizzard Visibility Table

Dice Roll	Distance
1 or 2	4"/10cm
3 or 4	8"/20cm
5 or 6	12"/30 cm